THE NANOTECHNOLOGY: THE FIFTH INDUSTRIAL REVOLUTION

DR. NASSER AFIFY

2019

Table of Contents

INTRODUCTION...	1
1- DEFINITION OF NANOTECHNOLOGY............	5
2- THE FOURTH INDUSTRIAL REVOLUTION........	18
3- THE SOCIETAL IMACTS OF NANOTECHNOLOGY	39
4- THE APPLICATIONS OF NANOTECHNOLOGY IN ENERGY TRANSMISSIONS	63
5- THE USES AND APPLICATIONS OF NANOTECHNOLOGY IN MEDICNE....... 100	
6- CONCLUSION...	121
REFERENCES.. 123	

INTRODUCTION

The fifth industrial revolution is the revolution of nanotechnology. Nanotechnology is science, engineering, and technology conducted at the nanoscale, which is about 1 to 100 nanometers. Nanoscience and nanotechnology are the study and application of extremely small things and can be used across all the other science fields, such as chemistry, biology, physics, materials science, and engineering.

The ideas and concepts behind nanoscience and nanotechnology started with a talk entitled "There's Plenty of Room at the Bottom" by physicist Richard Feynman at an American Physical Society meeting at the California Institute of Technology (CalTech) on December 29, 1959, long before the term nanotechnology was used. In his talk, Feynman described a process in which scientists would be able to manipulate and control individual atoms and molecules. Over a decade later, in his explorations of ultraprecision machining, Professor Norio Taniguchi coined the term nanotechnology. It wasn't until 1981, with the development of the scanning tunneling microscope that could "see" individual atoms that modern nanotechnology began.

It's hard to imagine just how small nanotechnology is. One nanometer is a billionth of a meter, or 10^{-9} of a meter. Here are a few illustrative examples: There are 25,400,000 nanometers in an inch. A sheet of newspaper is about 100,000 nanometers thick. On a comparative scale, if a marble were a nanometer, then one meter would be the size of the Earth.

Nanoscience and nanotechnology involve the ability to see and to control individual atoms and molecules. Everything on Earth is made up of atoms—the food we eat, the clothes we wear, the buildings and houses we live in, and our own bodies. But something as small as an atom is impossible to see with the naked eye. In fact, it's impossible to see with the microscopes typically used in a high school science classes. The microscopes needed to see things at the nanoscale were invented relatively recently—about 30 years ago. Once scientists had the right tools, such as the scanning tunneling microscope (STM) and the atomic force microscope (AFM), the age of nanotechnology was born.

Although modern nanoscience and nanotechnology are quite new, nanoscale materials were used for centuries. Alternate-sized gold and silver particles created colors in the stained glass windows of medieval churches hundreds of years ago. The artists back then just didn't know that the process they used to create these beautiful works of art actually led to changes in the composition of the materials they were working with.

Today's scientists and engineers are finding a wide variety of ways to deliberately make materials at the nanoscale to take advantage of their enhanced properties such as higher strength, lighter weight, increased control of light spectrum, and greater chemical reactivity than their larger-scale counterparts.

Medieval stained glass windows are an example of how nanotechnology was used in the pre-modern era. (Courtesy: NanoBioNet)

1. DEFINITION OF NANOTECHNOLOGY

Nanotechnology is the science of manipulating materials on an atomic or molecular scale especially to build microscopic devices (such as robots).

Nanotechnology, or nanotech for short, deals with matter at a level that most of us find hard to imagine, since it involves objects with dimensions of 100 billionths of a meter (1/800th of the thickness of a human hair) or less. The chemical and physical properties of materials often change greatly at this scale. Nanotechnology is already being used in automobile tires, land-mine detectors, and computer disk drives. Nanomedicine is a particularly exciting field: Imagine particles the size of a blood cell that could be released into the bloodstream to form into tiny robots and attack cancer cells, or "machines" the size of a molecule that could actually repair the damaged interiors of individual cells.

Also, it is the branch of technology that deals with dimensions and tolerances of less than 100 nanometres, especially the manipulation of individual atoms and molecules.

Because nanotechnology is still evolving, there doesn't seem to be any one definition that everybody agrees on. We know that nano deals with matter on a very small scale - larger than atoms and molecules, but smaller than a breadcrumb. We know that matter at the nano scale can behave differently than bulk matter. Beyond that, different individuals and groups focus on different aspects of nanotechnology as a discipline.

Nanotechnology is the study and use of structures between 1 nanometer (nm) and 100 nanometers in size .

This is probably the most barebones and generally agreed upon definition of nanotechnology. To put these measurements in perspective, compare your one meter (about three feet three inches) high hall table to a nanometer. You would have to stack one billion nanometer-sized particles on top of each other to reach the height of your hall table. Another popular comparison is that you can fit about 80,000 nanometers in the width of a single human hair.

It is related to structures, devices, and systems having novel properties and functions due to the arrangement of their atoms on the 1 to 100 nanometer scale. Many fields of endeavor contribute to nanotechnology, including molecular physics, materials science, chemistry, biology, computer science, electrical engineering, and mechanical engineering.

Nanotechnology is the study of phenomena and fine-tuning of materials at atomic, molecular and macromolecular scales, where properties differ significantly from those at a larger scale. Products based on nanotechnology are already in use and analysts expect markets to grow by hundreds of billions of euros during this decade.

Nanotechnology is the understanding and control of matter at dimensions between approximately 1 and 100 nanometers, where unique phenomena enable novel applications. Encompassing nanoscale science, engineering, and technology, nanotechnology involves imaging, measuring, modeling, and manipulating matter at this length scale.

In the context of computer science, it is a type of engineering geared toward building electronic components and devices measured in nanometers, which are extremely tiny in size and structure. Nanotechnology facilitates the building of functional matter and systems at the scalar level of an atom or molecule. It incorporates concepts from physics, biology, engineering and many other disciplines.

Nanotechnology is a scientific field that uses system or component development techniques to build products on highly granular levels. Nanotechnology works through different approaches to build nano materials or products, including bottom-up, top-down and functional system development. In a bottom-up approach, a product is designed as it evolves from its tiniest form factor to larger product. In a top-down approach, a large product may be reverse engineered to develop products scaled according to nanometer. A functional approach deals with a complete system and may incorporate bottom-up and top-down approaches.

Nanotechnology is implemented in many different fields and applications, such as computing, biotechnology, electronics and chemical engineering.

Nanotechnology is a field of research and innovation concerned with building 'things' - generally, materials and devices - on the scale of atoms and molecules. A nanometer is one-billionth of a metre: ten times the diameter of a hydrogen atom. The diameter of a human hair is, on average, 80,000 nanometres. At such scales, the ordinary rules of physics and chemistry no longer apply. For instance, materials' characteristics, such as their color, strength, conductivity and reactivity, can differ substantially between the nanoscale and the macro. Carbon 'nanotubes' are 100 times stronger than steel but six times lighter.

Nanotechnology is hailed as having the potential to increase the efficiency of energy consumption, help clean the environment, and solve major health problems. It is said to be able to massively increase manufacturing production at significantly reduced costs. Products of nanotechnology will be smaller, cheaper, lighter yet more functional and require less energy and fewer raw materials to manufacture, claim nanotech advocates.

In June 1999, Richard Smalley, Nobel laureate in chemistry, addressed the US House Committee on Science on the benefits of nanotechnology. "The impact of nanotechnology on the health, wealth, and lives of people," he said, "will be at least the equivalent of the combined influences of microelectronics, medical imaging, computer-aided engineering and man-made polymers developed in this century".

Others, however, are as cautious as Smalley is enthusiastic. Eric Drexler, the scientist who coined the term nanotechnology, has warned of developing "extremely powerful, extremely dangerous technologies". In his book Engines of Creation, Drexler envisioned that self-replicating molecules created by humans might escape our control. Although this theory has been widely discredited by researchers in the field, many concerns remain regarding the effects of nanotechnology on human and environmental health as well as the effect the new industry could have on the North-South divide. Activists worry that the science and development of nanotechnology will progress faster than policy-makers can devise appropriate regulatory measures. They say an informed debate must take place to determine the balance between risks and benefits.

Given the promise of nanotechnology, the race is on to harness its potential - and to profit from it. Many governments believe nanotechnology will bring about a new era of productivity and wealth, and this is reflected by the way public investment in nanotechnology research and development has risen during the past decade. In 2002, Japan was dedicating US$750 million a year to the field, a six-fold increase on the 1997 figure.

The US National Science Foundation predicts that the global market for nanotech-based products will exceed US$1 trillion within 15 years. Paul Miller, senior researcher at the British policy research organisation Demos, said in

2002 that "already, roughly one-third of the research budgets of the biggest science-based firms in the US is going into nanotech" whilst the US National Nanotechnology Initiative's budget rose from US$116 million in 1997 to a requested US$849 million in 2004.

In the developing world, Brazil, Chile, China, India, the Philippines, South Korea, South Africa and Thailand have shown their commitment to nanotechnology by establishing government-funded programmes and research institutes. Researchers at the University of Toronto Joint Centre for Bioethics have classified these countries as 'front-runners' (China, South Korea, India) and 'middle ground' players (Thailand, Philippines, South Africa, Brazil, Chile). In addition, Argentina and Mexico are 'up and comers': although they have research groups studying nanotechnology, their governments have not yet organized dedicated funding.

In May 2004, the Thai government announced plans to use nanotechnology in one per cent of all consumer products by 2013. Their market value by then is predicted to be 13 trillion baht (more than US$320 billion at contemporary exchange rates). Indeed, Thailand has wholeheartedly embraced nanotechnology and its development is a major commitment of the Thai government. Likewise, China announced in May 2004 that nanotechnology is central to its long-term national science and technology plan.

Nanotechnology holds the promise of new solutions to problems that hinder the development of poor countries, especially in relation to health and sanitation, food security, and the environment. In its 2005 report entitled Innovation: applying knowledge in development, the UN Millennium Project task force on science technology and innovation wrote that "nanotechnology is likely to be particularly important in the developing world, because it involves little labour, land or maintenance; it is highly productive and inexpensive; and it requires only modest amounts of materials and energy."

Nanotechnology is already useful as a tool in health care research. In January 2005, researchers at the US Massachusetts Institute of Technology used 'optical tweezers' - pairs of tiny glass beads are brought together or moved apart using laser beams - to study the elasticity of red blood cells that are infected with the malaria parasite (see Tiny tools tackle malaria). The technique is helping researchers to better understand how malaria spreads through the body.

But nanotechnology could also one day lead to cheaper, more reliable systems for drug-delivery. For example, materials that are built on the nanoscale can provide encapsulation systems that protect and secrete the enclosed drugs in a slow and controlled manner. This could be a valuable solution in countries that don't have adequate storage facilities and distribution networks, and for patients on complex drug regimens who cannot afford the time or money to travel long distances for a medical visit.

Tiny sensors offer the possibility of monitoring pathogens on crops and livestock as well as measuring crop productivity. In addition, nanoparticles could increase the efficiency of fertilizers. However, the Swiss insurance company SwissRe warned in a report in 2004 that they could also increase the ability of potentially toxic substances, such as fertilizers, to penetrate deep layers of the soil and travel over greater distances.

In addition, researchers in both developed and developing countries are developing crops that are able to grow under 'hostile' conditions, such as fields where the soil contains high levels of salt (sometimes due to climate change and rising sea levels) or low levels of water. They are doing this by manipulating the crops' genetic material, working on a nanotechnology scale with biological molecules. The application of nanotechnology in the field of renewable and sustainable energy (such as solar and fuel cells) could provide cleaner and cheaper sources of energy. These would improve both human and environmental health.

Tiny wastewater filters, for example, could sift emissions from industrial plants, eliminating even the smallest residues before they are released into the environment. Similar filters could clean up emissions from industrial combustion plants. And nanoparticles could be used to clean up oil spills, separating the oil from sand, removing it from rocks and from the feathers of birds caught in a spill.

Research has shown that nano-sized particles accumulate in the nasal cavities, lungs and brains of rats, and that carbon nanomaterials known as 'buckyballs' induce brain damage in fish. Vyvyan Howard, a toxicologist at the University of Liverpool in the United Kingdom, has warned that the small size of nanoparticles could render them toxic, and warns that full hazard assessments are needed before manufacturing is licensed.

Many interested parties, including the Canadian ETC Group and the insurance company SwissRe, have expressed their concern over releasing tiny particles which, because of their small size, are able to travel very far into the environment. They warn that we do not yet know how these particles will act in the environment or what chemical reactions they will trigger on meeting other particles. However, these same groups also concur with nanotechnology advocates who feel the field may offer 'cleaner' technologies, and, ultimately, a cleaner environment. But mostly, the concern is for the lack of research into nanotechnology's potential threats to human health, society and the environment.

Advances in nanotechnology have built on advances in microscopy. As well as allowing molecules to be imaged, the Scanning Tunnelling Microscope (patented in 1982) allowed researchers to manipulate them by picking up and moving individual atoms. This is the essence of 'bottom up' or molecular nanotechnology - the notion that molecular structures can be built atom-by-atom.

Some claim that nanotechnology could ultimately lead to the miniaturization of controlled production to the molecular level in much the same way as happens in human cells when, for instance, enzymes break and rearrange bonds holding molecules together. The vision is of potentially self-replicating 'assemblers' - tiny devices operating in unison like miniature versions of factory assembling lines - to produce 'nanomaterials', new products that will revolutionize construction, medicine, space exploration and computing.

The theory is well ahead of current realities and while some warn that self-replicating 'nanobots' pose an immense threat to humanity, others dismiss the idea as impossible. However, a recent production of a nano-conveyor belt that moves streams of particles rather than individual ones along a nanotube represents a major breakthrough, as does the development of a 'DNA robot' ten nanometres long capable of 'walking' along a pavement also made of DNA. Other significant developments are the discovery of spinning molecular structures, which herald the possibility of power generation and controllable motion at the molecular level.

Nanotech knowledge is rapidly growing. The number of scientific publications in the field grew from about 200 in 1997 to more than 12,000 in 2002. Despite this, relatively few products using nanoparticles are currently on the market. On the whole, the ones that are already on sale do not address the issues highlighted above, of health, food security and the environment. Rather, they have focused on consumer applications that include improved sunscreens, crack-resistant paints and scratch-proof spectacle lenses.

Like electricity and the internal combustion engine, nanotechnology is an enabling technology. As such, it is predicted to precipitate a range of innovations.

Assessing the role of nanotechnology and guiding its progression will require cross-sectorial involvement of scientists, governments, civil society organizations and the general public. Informed debate is essential to try to avoid the polarization of views illustrated by the issue of genetic modification. This 'quick guide' aims to provide a range of relevant information for those who would like to better understand and take part in this important debate.

Nanotechnology may be considered the most important feature of the fifth industrial revolution which following the ongoing fourth industrial one.

2. THE FOURTH INDUSTRIAL REVOLUTION

The fourth industrial revolution is the current and developing environment in which disruptive technologies and trends such as the Internet of Things (IoT), robotics, virtual reality (VR) and artificial intelligence (AI) are changing the way we live and work.

The third industrial revolution, sometimes called the digital revolution, involved the development of computers and IT (information technology) since the middle of the 20th century. The fourth industrial revolution is growing out of the third but is considered a new era rather than a continuation because of the explosiveness of its development and the disruptiveness of its technologies. According to Professor Klaus Schwab, Founder and Executive Chairman of the World Economic Forum and author of The Fourth Industrial Revolution, the new age is differentiated by the speed of technological breakthroughs, the pervasiveness of scope and the tremendous impact of new systems.

The Fourth Industrial Revolution heralds a series of social, political, cultural, and economic upheavals that will unfold over the 21st century. Building on the widespread availability of digital technologies that were the result of the Third Industrial, or Digital, Revolution, the Fourth Industrial Revolution will be driven largely by the convergence of digital, biological, and physical innovations.

Like the First Industrial Revolution's steam-powered factories, the Second Industrial Revolution's application of science to mass production and manufacturing, and the Third Industrial Revolution's start into digitization, the Fourth Industrial Revolution's technologies, such as artificial intelligence, genome editing, augmented reality, robotics, and 3-D printing, are rapidly changing the way humans create, exchange, and distribute value.

As occurred in the previous revolutions, this will profoundly transform institutions, industries, and individuals. More importantly, this revolution will be guided by the choices that people make today: the world in 50 to 100 years from now will owe a lot of its character to how we think about, invest in, and deploy these powerful new technologies.

It's important to appreciate that the Fourth Industrial Revolution involves a systemic change across many sectors and aspects of human life: the crosscutting impacts of emerging technologies are even more important than the exciting capabilities they represent. Our ability to edit the building blocks of life has recently been massively expanded by low-cost gene sequencing and techniques such as CRISPR; artificial intelligence is augmenting processes and skill in every industry; neurotechnology is making unprecedented strides in how we can use and influence the brain as the last frontier of human biology; automation is disrupting century-old transport and manufacturing paradigms; and technologies such as block chain and smart materials are redefining and blurring the boundary between the digital and physical worlds.

The result of all this is societal transformation at a global scale. By affecting the incentives, rules, and norms of economic life, it transforms how we communicate, learn, entertain ourselves, and relate to one another and how we understand ourselves as human beings.

Furthermore, the sense that new technologies are being developed and implemented at an increasingly rapid pace has an impact on human identities, communities, and political structures. As a result, our responsibilities to one another, our opportunities for self-realization, and our ability to positively impact the world are intricately tied to and shaped by how we engage with the technologies of the Fourth Industrial Revolution.

This revolution is not just happening to us—we are not its victims—but rather we have the opportunity and even responsibility to give it structure and purpose.

As economists Erik Brynjolfsson and Andrew McAfee have pointed out, this revolution could yield greater inequality, particularly in its potential to disrupt labor markets. As automation substitutes for labor across the entire economy, the net displacement of workers by machines might exacerbate the gap between returns to capital and returns to labor. On the other hand, it is also possible that the displacement of workers by technology will, in aggregate, result in a net increase in safe and rewarding jobs.

All previous industrial revolutions have had both positive and negative impacts on different stakeholders. Nations have become wealthier, and technologies have helped pull entire societies out of poverty, but the inability to fairly distribute the resulting benefits or anticipate externalities has resulted in global challenges.

By recognizing the risks, whether cybersecurity threats, misinformation on a massive scale through digital media, potential unemployment, or increasing social and income inequality, we can take the steps to align common human values with our technological progress and ensure that the

Fourth Industrial Revolution benefits human beings first and foremost.

We cannot foresee at this point which scenario is likely to emerge from this new revolution. However, I am convinced of one thing—that in the future, talent, more than capital, will represent the critical factor of production.

With these fundamental transformations underway today, we have the opportunity to proactively shape the Fourth Industrial Revolution to be both inclusive and human-centered. This revolution is about much more than technology—it is an opportunity to unite global communities, to build sustainable economies, to adapt and modernize governance models, to reduce material and social inequalities, and to commit to values-based leadership of emerging technologies.

The Fourth Industrial Revolution is therefore not a prediction of the future but a call to action. It is a vision for developing, diffusing, and governing technologies in ways that foster a more empowering, collaborative, and sustainable foundation for social and economic development, built around shared values of the common good, human dignity, and intergenerational stewardship. Realizing this vision will be the core challenge and great responsibility of the next 50 years.

Humanity continues to embark on a period of unparalleled technological advancement. The next 5, 10 and 20 years will present both significant challenges and opportunities. Private sectors, governments, academics and entrepreneurs are all seeking the roadmap for navigating these profound changes in the world of work. Such a road map must be created collaboratively by all stakeholders.

At its core, an industrial revolution can be characterized by advancements in technology that humanity applies to improve the process of production. But in reality, it means so much more.

The first three industrial revolutions brought to the world water and steam power, electricity and digitization. With every industrial revolution comes refining shifts to social, economic, environmental and political systems that truly alter the course of humanity. Some of these shifts are foreseen, and others are completely unforeseen.

Today, a fourth industrial revolution unfolds. The Fourth Industrial Revolution is bringing technologies that blur the lines between the physical, digital and biological spheres across all sectors. Technologies like artificial intelligence (AI), nanotechnology, quantum computing, synthetic biology and robotics will all drastically supersede any digital progress made in the past 60 years and create realities that we previously thought to be unthinkable. Such profound realities will disrupt and change the business model of each and every industry.

One of the most immediate and impactful outcomes of technological evolution is the vast advancement in automation. Every day, more manual process become automated, and as technology continues to accelerate, so will automation.

As a result, the world of work and labor market demand are rapidly changing. According to McKinsey, up to 375 million workers may need to change their occupational category by 2030, and digital work could contribute $2.7 trillion to global GDP by 2025. Faced with the scale of the unstoppable shifts in workforce demands, we must address the challenges associated with workforce transformation, starting by taking an in-depth look at its impact on the world of work. Four key impact areas should be considered:

For most global industries (e.g., logistics, financial, manufacturing, aerospace, etc.), advancements in AI, robotics, 3D printing and the internet of things will put a great deal of pressure on companies to automate in order to remain competitive in a global landscape. This will require companies to have a solid understanding of the way these technologies impact their industries and how they can ensure organizational agility to adapt to these changes. Increased global competitiveness will accelerate cost pressure, which will lead to substantial downsizing or reassignment of a large contingent of workers. McKinsey estimates that up to 800 million individuals may be displaced by automation by 2030.

There are four factors of production that fuel economic growth: land, labor, capital and enterprise. Today, the world is attaining only 52% of its entrepreneurial capacity, and this number is declining year over year. Large, established enterprises have a significant advantage in the future of work than smaller companies due to their ability to adapt to technological changes. However, this is not a recipe for long-term, sustainable economic success. The world must focus on supporting independent entrepreneurs, as small and midsize businesses are the fuel of most economies of the world today.

Technology will continue to change societal values. Today, more than 36% of the U.S. workforce are freelancers for reasons including autonomy, flexibility and extra income. Co-working spaces are exploding in popularity and are often fully subscribed before opening their doors. Technology has enabled people to work anytime, anywhere. By 2027, more than half of American workers will be freelancing.

Education and training: Part-and-parcel with economic development is one's ability to access training for employment. Naturally, tectonic shifts are happening in the education space. Students are less interested in stale curriculums and keener to take shorter, skills-based training that is more relevant to today's workplace. Employers are focusing on the skills required to achieve their business objectives and remain competitive and agile, which requires them to ensure their employees the necessary training to fill these skills gaps. Workers, naturally, need to acquire skills "on demand" to adapt to their changing roles and responsibilities.

Despite the challenges we face, we also possess an unprecedented possibility to apply an abundance mindset to solving the challenges. The Fourth Industrial Revolution will provide us with an opportunity to learn and teach new skills, build new jobs requiring unique skills combinations that don't exist today, explore talent that we didn't know about and, in doing so, grow our businesses and create a new generation of workers that are highly skilled in more diverse areas. The question is, how do we get there?

Collaborations among the private sector, academia and policymakers will be essential to navigate the future of work as we go through these profound moments. Schools need to work with businesses and the public sector to develop on-demand, relevant, adaptable curriculums and focus on teaching skills; governments need to utilize advanced technologies to generate real-time and predictive insights on the labor market in order to develop sound policies, programming and budgets; companies need to hire for competencies over credentials and, more importantly, take the lead in supporting existing workforces' upskilling and lifelong learning.

The Fourth Industrial Revolution is changing how we live, work, and communicate. It's reshaping government, education, healthcare, and commerce—almost every aspect of life. In the future, it can also change the things we value and the way we value them. It can change our relationships, our opportunities, and our identities as it changes the physical and virtual worlds we inhabit and even, in some cases, our bodies.

Education and access to information can improve the lives of billions of people. Through increasingly powerful computing devices and networks, digital services, and mobile devices, this can become a reality for people around the world, including those in underdeveloped countries.

The social media revolution embodied by Facebook, Twitter, and Tencent has given everyone a voice and a way to communicate instantly across the planet. Today, more than 30% of the people in the world use social media services to communicate and stay on top of world events.

These innovations can create a true global village, bringing billions more people into the global economy. They can bring access to products and services to entirely new markets. They can give people opportunities to learn and earn in new ways, and they can give people new identities as they see potential for themselves that wasn't previously available.

"The Fourth Industrial Revolution, finally, will change not only what we do but also who we are. It will affect our identity and all the issues associated with it: our sense of privacy, our notions of ownership, our consumption patterns, the time we devote to work and leisure, and how we develop our careers, cultivate our skills, meet people, and nurture relationships." —Klaus Schwab, The Fourth Industrial Revolution

Online shopping and delivery services—including by drone—are already redefining convenience and the retail experience. The ease of delivery can transform communities, even in remote places, and jumpstart the economies of small or rural areas.

In the physical realm, advances in biomedical sciences can lead to healthier lives and longer life spans. They can lead to innovations in neuroscience, like connecting the human brain to computers to enhance intelligence or experience a simulated world. Imagine all that robot power with human problem-solving skills.

Advances in automotive safety through Fourth Industrial Revolution technologies can reduce road fatalities and insurance costs, and carbon emissions. Autonomous vehicles can reshape the living spaces of cities, architecture, and roads themselves, and free up space for more social and human-centered spaces.

Digital technology can liberate workers from automatable tasks, freeing them to concentrate on addressing more complex business issues and giving them more autonomy. It can also provide workers with radically new tools and insights to design more creative solutions to previously insurmountable problems.

However, while the Fourth Industrial Revolution has the power to change the world positively, we have to be aware that the technologies can have negative results if we don't think about how they can change us.

We build what we value. This means we need to remember our values as we're building with these new technologies. For example, if we value money over family time, we can build technologies that help us make money at the expense of family time. In turn, these technologies can create incentives that make it harder to change that underlying value.

People have a deep relationship with technologies. They are how we create our world, and we have to develop them with care. More than ever, it's important that we begin right. We have to win this race between the growing power of the technology, and the growing wisdom with which we manage it. We don't want to learn from mistakes.

Biotechnology can lead to controversial advances such as designer babies, gene drives (changing the inherited traits of an entire species), or implants required to become competitive candidates for schools or jobs. Innovations in robotics and automation can lead to lost jobs, or at least jobs that are very different and value different skills.

Artificial intelligence, robotics, bioengineering, programming tools, and other technologies can all be used to create and deploy weapons.

Social media can erase borders and bring people together, but it also can also intensify the social divide. And it gives voice to cyber-bullying, hate speech, and spreading false stories. We have to decide what kind of social media rules we want to create, but we also have to accept that social media is reshaping what we value and how we create and deploy those rules.

In addition, being always connected can turn into a liability, with no respite from the continuous overload of data and connections.

Artificial intelligence is unleashing a whole new level of productivity and augmenting our lives in many ways. As in past industrial revolutions, it can also be a disruptive force, dislocating people from jobs and surfacing questions about the relationship between humans and machines.

It's inevitable that jobs are going to be impacted as artificial intelligence automates a variety of tasks. However, just as the Internet did 20 years ago, the artificial intelligence revolution is going to transform many jobs— and spawn new kinds of jobs that drive economic growth. Workers can spend more time on creative, collaborative, and complex problem-solving tasks that machine automation isn't well suited to handle.

However, workers with less education and fewer skills are at a disadvantage as the Fourth Industrial Revolution progresses. Businesses and governments need to adapt to the changing nature of work by focusing on training people for the jobs of tomorrow. Talent development, lifelong learning, and career reinvention are going to be critical to the future workforce.

People are asking whether the Fourth Industrial Revolution is the road to a better future for all. The power of technology is increasing rapidly and facilitating extraordinary levels of innovation. And as we know, more people and things in the world are becoming connected. But that doesn't necessarily pave the way for a more open, diverse, and inclusive global society.

The lessons of previous industrial revolutions include the realization that technology and its wealth generation can serve the interests of small, powerful groups above the rest. Powerful new technologies built on global digital networks can be used to keep societies under undue surveillance while making us vulnerable to physical and cyberattacks. These are the challenges we can face to make sure the combination of technology and politics together don't create disparities that hinder people.

According to the World Economic Forum Global Risks Report 2017, "the Fourth Industrial Revolution has the potential to raise income levels and improve the quality of life for all people. But today, the economic benefits of the Fourth Industrial Revolution are becoming more concentrated among a small group. This increasing inequality can lead to political polarization, social fragmentation, and lack of trust in institutions. To address these challenges, leaders in the public and private sectors need to have a deeper commitment to more inclusive development and equitable growth that lifts up all people".

Many people around the world haven't yet benefited from previous industrial revolutions. As the authors of Shaping the Fourth Industrial Revolution point out, at least 600 million people live on smallholder farms without access to any mechanization, living lives largely untouched by the first industrial revolution. Around one-third of the world's population (2.4 billion) lack clean drinking water and safe sanitation, around one-sixth (1.2 billion) have no electricity—both systems developed in the second industrial revolution. And while the digital revolution means that more than 3 billion people now have access to the Internet, that still leaves more than 4 billion out of a core aspect of the third industrial revolution.

The means that as we appreciate and engage with the exciting technologies of the Fourth Industrial Revolution, we must work to ensure that the opportunities they bring are well-distributed around the world and across our communities. In particular, we must help those who missed out on the huge increases in quality of life that the first, second, and third industrial revolutions provided.

"Let us together shape a future that works for all by putting people first, empowering them and constantly reminding ourselves that all of these new technologies are first and foremost tools made by people for people." — Klaus Schwab, The Fourth Industrial Revolution.

We value the ability to control what is known about us, and yet we are living in a world where tracking every individual's personal information is key to delivering more intelligent, personalized services. For example:

Facebook tracks what you do so that it knows which content and advertisements are most relevant to you.

Smartphones track your location, and you can share that information with apps that recommend places to eat or shop. Retailers analyze your purchase history to recommend products and offer coupons to stimulate more sales.

In the future, you'll walk into a store and the salesperson will immediately have your name, credit rating, marital status, and past purchases flashed to their augmented-reality virtual screen.

Technological advances are also broadening the scope of surveillance. In the UK today, an estimated 6 million CCTV cameras are recording activity all over the country. Advances in computing power and artificial intelligence can potentially enable law enforcement agencies to track suspected terrorists by analyzing social networks, government records, and other data.

In the future, billions of 3D-printed "smart dust" cameras floating in the air can monitor the activities of humans. From traffic reports to natural disasters, such technology can keep us safer. But it also can watch us when we do not want to be watched.

For consumers, businesses that are transparent about their data collection practices and that prioritize consumer privacy can win our loyalty.

Public trust in business, government, the media, and even technology is falling. This is a crisis that is dividing societies and creating instability around the world.

The technologies of the Fourth Industrial Revolution themselves are neutral, but are they being applied in ways that build trust? Are consumers going to trust that new artificial intelligence and robotic systems can make their lives better, or are they going to be fearful of the machines and those who control them? Are citizens going to trust the institutions and service providers who collect and maintain their data?

For the Fourth Industrial Revolution to generate trust, everyone contributing to it (including you) must collaborate and feel a connection to common objectives. More transparency into how we govern and manage this technology is key, as are security models that boost our confidence that these systems won't be hacked, run amok, or become tools of oppression by those who control them.

The innovations in artificial intelligence, biotechnology, robotics, and other emerging technologies are going to redefine what it means to be human and how we engage with one another and the planet. Our capabilities, our identities, and our potential will all evolve along with the technologies we create.

In the coming decades, we must establish guardrails that keep the advances of the Fourth Industrial Revolution on a track to benefit all of humanity. We must recognize and manage the potential negative impacts they can have, especially in the areas of equality, employment, privacy, and trust. We have to consciously build positive values into the technologies we create, think about how they are to be used, and design them with ethical application in mind and in support of collaborative ways of preserving what's important to us.

This effort requires all stakeholders—governments, policymakers, international organizations, regulators, business organizations, academia, and civil society—to work together to steer the powerful emerging technologies in ways that limit risk and create a world that aligns with common goals for the future.

You, as a person, citizen, employee, investor, and social influencer, are a critical stakeholder in the Fourth Industrial Revolution. Sharing your thoughts on the technologies and what you value as this revolution unfolds is essential. The world we create through technologies can shape our lives and is the one we pass on to the next generation.

"The Fourth Industrial Revolution can compromise humanity's traditional sources of meaning—work, community, family, and identity—or it can lift humanity into a new collective and moral consciousness based on a sense of shared destiny. The choice is ours." —Klaus Schwab, The Fourth Industrial Revolution.

3. THE SOCIETAL IMACTS OF NANOTECHNOLOGY

Societal impact is how institutions, organization, businesses or individuals actions affect the surrounding society. The social implications of any new technology can be felt by people directly incorporated with organization or individual or people in different societies and countries. The societal impacts of new technologies are easy to identify but hard to measure or predict. Nanotechnology will have significant social impacts in the areas of military applications, intellectual property issues, as well as having an effect on labor and the balance between citizens and governments.

A high proportion of nanotechnology research is sponsored by the military and thus focused towards military applications. The potential military applications include nanorobotics, magnetorheological fluid (MRF), artificial intelligence and molecular manufacturing.

The advanced developments in the military technology may have implications for societal and political relations within the community. Modern defense armies are protected from today's civilian threats in a way that never had before. It is likely that nanotechnology will further widen the gap between the means of political violence available to the military and those available to the civilian population.

The advancement of nanotechnology might contribute to terrorism, as it can exacerbate existing trends towards asymmetric warfare. If the military forces of the industrialized world become more difficult to attack and defeat due to nanotechnology, this may force those involved in the war with them to adopt new strategies, including sabotage and attacks on civilian and other targets. The enhancement of military application in nanotechnology may thus indirectly increase the occurrence of terrorist attacks in the future.

The patent attorney must establish uniqueness and obviousness in the process of nanotechnology patent application. A patent examiner may state that a nanostructured product lacks novelty because the relevant nanostructure material was present in an existing product, even though the nanostructure material was not recognized.

Pundits have warned that the resulting patent creates an adverse effect in progress in technology and have argued that there should be held patents on "basic" nanotechnologies. IBM hold an early and basic patent on single-wall CNT which can identify as one of the most significant patents that could have an impact on the future development of nanotechnology. CNT have great potential to replace major conventional raw materials. However, as their application expands, anyone manufactures or sell CNT, no matter what the uses, must first buy a license from IBM.

The nanotechnology impact on labor is in its use of particular factors of production. During the improvement of nanotechnology, firms are likely to have high demands for the scientists, engineers, and technicians who have to build and integrate the new ideas into processes and products. In addition, there is a need for supporting labor services, which creates career opportunities.

Nanotechnology is likely to have even less impact on labor market inequalities. This is because most of us not need literate in nanotechnologies any more than we are literate about computer circuit design. The nanotechnology can be expected to concentrate political power in the hands of governments. Nanotechnology can be expected to be applied to further miniaturize and advance surveillance technologies such as cameras, listening devices, tracking devices, and face and pattern recognition systems.

The improvements in the field of electronics and computer memory that nanotechnology makes possible capacity of government to collect, store, and examine data. Developments in nanotechnology, can therefore, be expected to increase significantly the ability of governments to keep track of their citizens.

The societal impact of nanotechnology is the potential benefits and challenges that the introduction of novel nanotechnological devices and materials may hold for society and human interaction. The term is sometimes expanded to also include nanotechnology's health and environmental impact, but this article will only consider the social and political impact of nanotechnology. As nanotechnology is an emerging field and most of its applications are still speculative, there is much debate about what positive and negative effects that nanotechnology might have.

Beyond the toxicity risks to human health and the environment which are associated with first-generation nanomaterials, nanotechnology has broader societal implications and poses broader social challenges. Social scientists have suggested that nanotechnology's social issues should be understood and assessed not simply as "downstream" risks or impacts. Rather, the challenges should be factored into "upstream" research and decision making in order to ensure technology development that meets social objectives. Many social scientists and organizations in civil society suggest that technology assessment and governance should also involve public participation.

Some observers suggest that nanotechnology will build incrementally, as did the 18-19th century industrial revolution, until it gathers pace to drive a nanotechnological revolution that will radically reshape our economies, our labor markets, international trade, international relations, social structures, civil liberties, our relationship with the natural world and even what we understand to be human. Others suggest that it may be more accurate to describe change driven by nanotechnology as a "technological tsunami". Just like a tsunami, analysts warn that rapid nanotechnology-driven change will necessarily have profound disruptive impacts. As the APEC Center for Technology Foresight observes:

If nanotechnology is going to revolutionize manufacturing, health care, energy supply, communications and probably defense, then it will transform labour and the workplace, the medical system, the transportation and power infrastructures and the military. None of these latter will be changed without significant social disruption.

Those concerned with the negative impact of nanotechnology suggest that it will simply exacerbate problems stemming from existing socio-economic inequity and unequal distributions of power, creating greater inequities between rich and poor through an inevitable nano-divide (the gap between those who control the new nanotechnologies and those whose products, services or labour are displaced by them).

Analysts suggest the possibility that nanotechnology has the potential to destabilize international relations through a nano arms race and the increased potential for bioweaponry; thus, providing the tools for ubiquitous surveillance with significant implications for civil liberties. Also, many critics believe it might break down the barriers between life and non-life through nanobiotechnology, redefining even what it means to be human.

Nanoethicists posit that such a transformative technology could exacerbate the divisions of rich and poor – the so-called "nano divide." However nanotechnology makes the production of technology, e.g. computers, cellular phones, health technology etcetera, cheaper and therefore accessible to the poor.

In fact, many of the most enthusiastic proponents of nanotechnology, such as transhumanists, see the nascent science as a mechanism to changing human nature itself – going beyond curing disease and enhancing human characteristics. Discussions on nanoethics have been hosted by the federal government, especially in the context of "converging technologies" – a catch-phrase used to refer to nano, biotech, information technology, and cognitive science.

Possible military applications of nanotechnology have been suggested in the fields of soldier enhancement and chemical weapons amongst others. However, more socially disruptive weapon systems are to be expected from molecular manufacturing, a potential future form of nanotechnology that would make it possible to build complex structures at atomic precision.

Molecular manufacturing requires significant advances in nanotechnology, but its supporters posit that once achieved it could produce highly advanced products at low costs and in large quantities in nanofactories weighing a kilogram or more. If nanofactories gain the ability to produce other nanofactories production may only be limited by relatively abundant factors such as input materials, energy and software.

Molecular manufacturing might be used to cheaply produce, among many other products, highly advanced, durable weapons. Being equipped with compact computers and motors these might be increasingly autonomous and have a large range of capabilities. According to Chris Phoenix and Mike Treder from the Center for Responsible Nanotechnology as well as Anders Sandberg from the Future of Humanity Institute the military uses of molecular manufacturing are the applications of nanotechnology that pose the most significant global catastrophic risk.

Several nanotechnology researchers state that the bulk of risk from nanotechnology comes from the potential to lead to war, arms races and destructive global government. Several reasons have been suggested why the availability of nanotech weaponry may with significant likelihood lead to unstable arms races (compared to e.g. nuclear arms races): (1) A large number of players may be tempted to enter the race since the threshold for doing so is low; (2) the ability to make weapons with molecular manufacturing might be cheap and easy to hide; (3) therefore lack of insight into the other parties' capabilities can tempt players to arm out of caution or to launch preemptive strikes; (4) molecular manufacturing may reduce dependency on international trade, a potential peace-promoting factor; (5) wars of aggression may pose a smaller economic threat to the aggressor since manufacturing is cheap and humans may not be needed on the battlefield.

Self-regulation by all state and non-state actors has been called hard to achieve, so measures to mitigate war-related risks have mainly been proposed in the area of international cooperation. International infrastructure may be expanded giving more sovereignty to the international level. This could help coordinate efforts for arms control. Some have put forth that international institutions dedicated specifically to nanotechnology (perhaps analogously to the International Atomic Energy Agency IAEA) or general arms control may also be designed.

One may also jointly make differential technological progress on defensive technologies. The Center for Responsible Nanotechnology also suggests some technical restrictions. Improved transparency regarding technological capabilities may be another important facilitator for arms-control.

On the structural level, critics of nanotechnology point to a new world of ownership and corporate control opened up by nanotechnology. The claim is that, just as biotechnology's ability to manipulate genes went hand in hand with the patenting of life, so too nanotechnology's ability to manipulate molecules has led to the patenting of matter. The last few years has seen a gold rush to claim patents at the nanoscale. Academics have warned that the resultant patent thicket is harming progress in the technology and have argued in the top journal Nature that there should be a moratorium on patents on "building block" nanotechnologies.

Over 800 nano-related patents were granted in 2003, and the numbers are increasing year to year. Corporations are already taking out broad-ranging patents on nanoscale discoveries and inventions. For example, two corporations, NEC and IBM, hold the basic patents on carbon nanotubes, one of the current cornerstones of nanotechnology. Carbon nanotubes have a wide range of uses, and look set to become crucial to several industries from electronics and computers, to strengthened materials to drug delivery and diagnostics. Carbon nanotubes are poised to become a major traded commodity with the potential to replace major conventional raw materials. However, as their use expands, anyone seeking to (legally) manufacture or sell carbon nanotubes, no matter what the application, must first buy a license from NEC or IBM.

Nanotechnologies may provide new solutions for the millions of people in developing countries who lack access to basic services, such as safe water, reliable energy, health care, and education. The United Nations has set Millennium Development Goals for meeting these needs. The 2004 UN Task Force on Science, Technology and Innovation noted that some of the advantages of nanotechnology include production using little labor, land, or maintenance, high productivity, low cost, and modest requirements for materials and energy.

Many developing countries, for example Costa Rica, Chile, Bangladesh, Thailand, and Malaysia, are investing considerable resources in research and development of nanotechnologies. Emerging economies such as Brazil, China, India and South Africa are spending millions of US dollars annually on R&D, and are rapidly increasing their scientific output as demonstrated by their increasing numbers of publications in peer-reviewed scientific publications.

Potential opportunities of nanotechnologies to help address critical international development priorities include improved water purification systems, energy systems, medicine and pharmaceuticals, food production and nutrition, and information and communications technologies. Nanotechnologies are already incorporated in products that are on the market. Other nanotechnologies are still in the research phase, while others are concepts that are years or decades away from development.

Applying nanotechnologies in developing countries raises similar questions about the environmental, health, and societal risks described in the previous section. Additional challenges have been raised regarding the linkages between nanotechnology and development.

Protection of the environment, human health and worker safety in developing countries often suffers from a combination of factors that can include but are not limited to lack of robust environmental, human health, and worker safety regulations; poorly or unenforced regulation which is linked to a lack of physical (e.g., equipment) and human capacity (i.e., properly trained regulatory staff). Often, these nations require assistance, particularly financial assistance, to develop the scientific and institutional capacity to adequately assess and manage risks, including the necessary infrastructure such as laboratories and technology for detection.

Very little is known about the risks and broader impacts of nanotechnology. At a time of great uncertainty over the impacts of nanotechnology it will be challenging for governments, companies, civil society organizations, and the general public in developing countries, as in developed countries, to make decisions about the governance of nanotechnology.

Companies, and to a lesser extent governments and universities, are receiving patents on nanotechnology. The rapid increase in patenting of nanotechnology is illustrated by the fact that in the US, there were 500 nanotechnology patent applications in 1998 and 1,300 in 2000. Some patents are very broadly defined, which has raised concern among some groups that the rush to patent could slow innovation and drive up costs of products, thus reducing the potential for innovations that could benefit low income populations in developing countries.

There is a clear link between commodities and poverty. Many least developed countries are dependent on a few commodities for employment, government revenue, and export earnings. Many applications of nanotechnology are being developed that could impact global demand for specific commodities. For instance, certain nanoscale materials could enhance the strength and durability of rubber, which might eventually lead to a decrease in demand for natural rubber.

Other nanotechnology applications may result in increases in demand for certain commodities. For example, demand for titanium may increase as a result of new uses for nanoscale titanium oxides, such as titanium dioxide nanotubes that can be used to produce and store hydrogen for use as fuel. Various organizations have called for international dialogue on mechanisms that will allow developing countries to anticipate and proactively adjust to these changes.

In 2003, Meridian Institute began the Global Dialogue on Nanotechnology and the Poor: Opportunities and Risks (GDNP) to raise awareness of the opportunities and risks of nanotechnology for developing countries, close the gaps within and between sectors of society to catalyze actions that address specific opportunities and risks of nanotechnology for developing countries, and identify ways that science and technology can play an appropriate role in the development process.

The GDNP has released several publicly accessible papers on nanotechnology and development, including "Nanotechnology and the Poor: Opportunities and Risks - Closing the Gaps Within and Between Sectors of Society"; "Nanotechnology, Water, and Development"; and "Overview and Comparison of Conventional and Nano-Based Water Treatment Technologies".

Concerns are frequently raised that the claimed benefits of nanotechnology will not be evenly distributed, and that any benefits (including technical and/or economic) associated with nanotechnology will only reach affluent nations. The majority of nanotechnology research and development - and patents for nanomaterials and products - is concentrated in developed countries (including the United States, Japan, Germany, Canada and France). In addition, most patents related to nanotechnology are concentrated amongst few multinational corporations, including IBM, Micron Technologies, Advanced Micro Devices and Intel.

This has led to fears that it will be unlikely that developing countries will have access to the infrastructure, funding and human resources required to support nanotechnology research and development, and that this is likely to exacerbate such inequalities.

Producers in developing countries could also be disadvantaged by the replacement of natural products (including rubber, cotton, coffee and tea) by developments in nanotechnology. These natural products are important export crops for developing countries, and many farmers' livelihoods depend on them. It has been argued that their substitution with industrial nano-products could negatively impact the economies of developing countries, that have traditionally relied on these export crops.

It is proposed that nanotechnology can only be effective in alleviating poverty and aid development "when adapted to social, cultural and local institutional contexts, and chosen and designed with the active participation by citizens right from the commencement point" (Invernizzi et al. 2008, p. 132).

Ray Kurzweil has speculated in The Singularity is Near that people who work in unskilled labor jobs for a livelihood may become the first human workers to be displaced by the constant use of nanotechnology in the workplace, noting that layoffs often affect the jobs based around the lowest technology level before attacking jobs with the highest technology level possible.

It has been noted that every major economic era has stimulated a global revolution both in the kinds of jobs that are available to people and the kind of training they need to achieve these jobs, and there is concern that the world's educational systems have lagged behind in preparing students for the "Nanotech Age".

It has also been speculated that nanotechnology may give rise to nanofactories which may have superior capabilities to conventional factories due to their small carbon and physical footprint on the global and regional environment.

The miniaturization and transformation of the multi-acre conventional factory into the nanofactory may not interfere with their ability to deliver a high quality product; the product may be of even greater quality due to the lack of human errors in the production stages. Nanofactory systems may use precise atomic precisioning and contribute to making superior quality products that the "bulk chemistry" method used in 20th century and early 21st currently cannot produce. These advances might shift the computerized workforce in an even more complex direction, requiring skills in genetics, nanotechnology, and robotics.

Nanotechnology will have broad applications across all fields of engineering, so it will be an amplifier of the social effects of other technologies. There is an especially great potential for it to combine with three other powerful trends – biotechnology, information technology, and cognitive science – based on the material unity of nature at the nanoscale and on technology integration from that scale.

It will be important to integrate social and ethical studies into nanotechnology developments from their very beginning. Technically competent research on the societal implications of nanotechnology will help give policymakers and the general public a realistic picture free of unreasonable hopes or fears. Nanotechnology is a booming technology that swiftly has entered society. Amongst the many nanotechnological products already available on the market are - besides technological devices in cars, computers and the like – food, health and beauty products. Nanotechnology as a term has not been very prominent in public discourse, although its connotation is rather positive.

In 2002, the US National Nanotechnology Initiative awarded only $280,000 — 0.04% of its budget of $697 million, to study the social and ethical implications of nanotechnology. None of this money was allocated to studying risk perception . However, knowledge worldwide is not yet substantiated enough to permit statements about health-related or environmental impacts of nanotechnological products. We lack reliable data and possible risks and need more in-depth (and in particular long-term) investigations into environmental and health impacts.

A US report named "Nanotechnology in agriculture and food production: anticipated applications", for the first time analyzes the publicly available data on federally funded research projects in agrifood nanotechnology, supplemented with data from the U.S. Patent and Trademark Office. Written by Jennifer Kuzma and Peter VerHage from the University of Minnesota's Center for Science, Technology, and Public Policy, the report estimates possible areas and timeframes for future nanotechnology-based food and agriculture applications. It takes an early look at potential benefits and risks, and it explores possible areas and needs for environmental, health and safety oversight.

After sociologist Etzkowitz the social sciences can play three different but mutually supportive roles in the development of nanotechnology: 1. Analyzing and contributing to the improvement of the processes of scientific discoveries that increasingly involve organizational issues where the social sciences have a long-term research and knowledge base. 2. Analyzing the effects of nanotechnology, whether positive or negative, expected or unintended, hypothetically and proactively and as they occur in realtime. 3. Evaluation of public and private programs to promote nanoscience and nanotechnology.

In the year 2000, the US National Science and Technology Council sponsored a major workshop at the

National Science Foundation, which led to a published report, Societal Implications of Nanoscience and Nanotechnology. About the involvement of social scientists in nanotechnology, it says "It is important to include a wide range of interests, values, and perspectives in the overall decision process that charts the future development of nanotechnology.

Involvement of members of the public or their representatives has the added benefit of respecting their interests and enlisting their support. The inclusion of social scientists and humanistic scholars, such as philosophers of ethics, in the social process of setting visions for nanotechnology is an important step for the National Nanotechnology Initiative. As scientists or dedicated scholars in their own right, they can respect the professional integrity of nanoscientists and nanotechnologists, while contributing a fresh perspective. Given appropriate support, they could inform themselves deeply enough about a particular nanotechnology to have a well-grounded evaluation.

At the same time, they are professionally trained representatives of the public interest and capable of functioning as communicators between nanotechnologists and the public or government officials. Their input may help maximize the societal benefits of the technology while reducing the possibility of debilitating public controversies.

Roblegg and coworkers from the Nanonet Styria, an Austrian Nanotechnology Network, published a report on health risks of nanotechnology. This report stresses the need for long-term studies on health implications of nanotechnology. Oberdörster et al. showed in animal experiments that there is translocation of inhaled ultrafine particles (smaller than 100nm) along the olfactory nerve into the olfactory bulb in the brain

. The significance of this study for humans, however, still needs to be established. The translocation of particles along nerve fibers could provide a portal of entry into the central nervous system for solid ultrafine particles, circumventing the tight blood–brain barrier. Whether this translocation of inhaled ultrafine particles can cause central nervous system effects needs to be determined in future studies. There are currently no studies on the behaviour of nanoparticles in cosmetics products.

Nanoparticles are for example found in sunscreen products and in skin creams. Long term studies are necessary, since currently, the US Food and Drug Administration, the FDA, as well as the Scientific Committee on Cosmetic Products and Non-Food Products intended for Consumers of the European Commission regard nanoparticles in cosmetics as a variation of the bulk material, ignoring possible non-scalable size effects. There is currently to need to perform time consuming and expensive toxicological tests. There was a similar situation regarding chiral pharmaceuticals. Left- and right-handed isomers (enantionmers) of the same molecule used to be regarded by the FDA as the same component.

An impressive (and tragic) example on how different the enantiomers of the same molecule can act in the human body was given by the substance Thalidomide. Thalidomide was sold in some countries under the name Contergan. One enantiomer of Thalidomide is effective against morning sickness (this is why it was administered to pregnant women). The other enantiomer is teratogenic, and causes birth defects (approximately 10 000 "Contergan babies" were born in the 1950s and 1960s).

The enantiomers are converted to each other in vivo – that is, if a human is given (R)-thalidomide or (S)-thalidomide, both isomers can be found in the serum – therefore, administering only one enantiomer will not prevent the teratogenic effect in humans. At the end of the 1990s a paradigm shift took place in the FDA and today, leftand right-handed isomers of pharmaceuticals are treated as two different substances.

Ethical questions related to nanotechnology are not limited to the ways people might use it to harm others intentionally, but also include obligations to avoid potentially harmful unintended consequences.

The best way to reassert the truth-oriented professional norms of science would be to rebuild good channels of communication and cooperation, reattaching the researchers to each other and to the scientific community. In a special issue on "Nanotech Challenges" of the HYLE International Journal for Philosophy of Chemistry, Lewenstein attempts to answer what counts as an ethical issue in nanotechnology.

He concludes that the attempts to define ethical issues narrowly is itself an exercise of power that can prevent us from understanding how central ethical issues are to the development of scientific knowledge and its implementation through technology in the modern world.

Nanotechnology's highest and best use should be to create a world of abundance where no one is lacking for their basic needs. Those needs include adequate food, safe water, a clean environment, housing, medical care, education, public safety, fair labour, unrestricted travel, artistic expression and freedom from fear and oppression. High priority must be given to the efficient and economical global distribution of the products and services created by nanotechnology.

Increased spending on pertinent research has resulted in the establishment of broad technological expertise and in a substantial number of important projects in the field of nanotechnology. However, currently there is a lack of capacity with regard to aspects of risk and healthrelated, environmental and societal implications of nanotechnology. In accordance with the Institute of Technology Assessment of the Austrian Academy of Sciences we propose to earmark a certain part (minimum 5%, as a guideline) of the special funding for nanotechnology for risk research and accompanying measures. Ultimately, the test of the various nanotechnologies will be their benefit for human beings, as measured by economic growth, improved health and longevity, environmental protection, strengthened security, social vitality, and enhanced human capabilities.

4. THE APPLICATIONS OF NANOTECHNOLOGY IN ENERGY TRANSMISSIONS

The application of nanotechnologies to energy transmission has the potential to significantly impact both the deployed transmission technologies and the need for additional development. This could be a factor in assessing environmental impacts of right-of-way (ROW) development and use. For example, some nanotechnology applications may produce materials (e.g., cables) that are much stronger per unit volume than existing materials, enabling reduced footprints for construction and maintenance of electricity transmission lines. Other applications, such as more efficient lighting, lighter-weight materials for vehicle construction, and smaller batteries having greater storage capacities may reduce the need for long-distance transport of energy, and possibly reduce the need for extensive future ROW development and many attendant environmental impacts.

Nanoscale materials (or nanomaterials) contain nanoparticles or are developed using nanotechnology. Nanoparticles are commonly considered to be materials that have at least one dimension that is less than 100 nm.

Nanoparticles can be distinguished according to their origin. They can occur naturally (e.g., from volcanic eruptions); they can be produced incidentally during other processes (e.g., from fuel combustion); and they can be manufactured intentionally. Manufactured nanomaterials can be classified according to their method of production. Some are produced "from the top down," as when a bulk material (e.g., gold, silicate) is reduced to a mass of nanoscale particles.

Because of their very small size, these nanoscale metals, metal oxides, powders, and dusts have physical, chemical, magnetic, electrical, mechanical, and other properties that differ from those of the bulk materials from which they are derived. They are found today in sunscreen products (titanium dioxide nanoparticles), solar cells (aluminum oxide nanoparticles), and several other applications in research and commerce.

The second type of manufactured nanoparticles are built "from the bottom up," atom-by atom or molecule-by-molecule. Engineered nanoparticles of this type are still relatively difficult and expensive to manufacture, but they have the potential to impact energy development and use, transportation, electronics, manufacturing, and other disciplines.

Nanoparticles and nanomanufacturing techniques may impact energy transmission system development and use for many years to come. For example, nanotechnologies may make more efficient use of transportation fuels, possibly slowing the increase in demand for long-distance shipment of liquid fuels. Construction materials made from nanoparticles may be stronger but occupy less volume than today's materials, which may reduce the footprints required for the construction and maintenance of pipelines and electricity transmission lines.

Nanotechnology is being used or considered for use in many applications targeted to provide cleaner, more efficient energy supplies and uses. While many of these applications may not affect energy transmission directly, each has the potential to reduce the need for the electricity, petroleum distillate fuel, or natural gas that would otherwise be moved through energy transmission ROWs. More efficient energy generation and use (and the consequent reduced need to transmit energy over long distances) may decrease the amount of construction, maintenance, repair, and decommissioning activities along the ROWs that would otherwise be needed to meet increased energy demands. Energy-related technologies in which nanotechnology may play a role include: • Lighting, • Heating, • Transportation, • Renewable energy, • Energy storage, • Fuel cells, • Hydrogen generation and storage, and • Power Chips.

Lighting in the United States, roughly 20% of all electricity is consumed in providing incandescent and fluorescent lighting. Because of their compactness, durability, low heat generation, and electrical efficiency, light-emitting diodes (LEDs) now rival incandescent light sources in many 6 parts of the visible spectrum and are being used in displays, automobile lights, and traffic lights. Semiconductors used in the preparation of LEDs for lighting are increasingly being built at nanoscale dimensions, and projections indicate that nanotechnology-based lighting advances have the potential to reduce worldwide consumption of energy by more than 10% (NNI 2000).

Nanocrystals, also known as quantum dots, are known primarily for their ability to produce distinct colors of light as the size of the individual crystals is varied. In 2005, researchers at Vanderbilt University coated an LED with a thin layer of quantum dots, thereby producing a hybrid LED that yielded warm, white light similar to that of an incandescent lamp. The discovery has implications for using nanotechnology to produce light for residential, commercial, and industrial applications without the heat that accounts for a large portion of the incandescent light bulb's poor energy efficiency (Salisbury 2005).

Heating Nanotechnology may help accelerate the development of energy-efficient central heating. When added to water, CNTs disperse to form a nanofluid. Researchers have developed nanofluids whose rates of forced convective heat transfer are four times better than the norm by using CNTs. When added to a home's commercial water boiler, such nanofluids could make the central heating device 10% more efficient. The researchers say that the technology is 3 to 5 years away from commercial home use (Pollitt 2006).

Transportation Nanotechnology may enable more efficient transportation via catalysts in fuels; lighter, stronger materials; and more efficient batteries.

The Envirox Fuel Borne Catalyst, developed by Oxonica, Ltd., is an example of a commercially proven product that improves diesel fuel combustion, reducing fuel consumption and harmful exhaust emissions. The additive uses nanoscale (10 nm across) particles of cerium oxide to catalyze the combustion reactions between diesel fuel and air. The small particle size creates a greater surface area for catalyzing the reactions, causes the particles to remain more evenly suspended in the fuel, and allows the additive to be used at concentrations as low as five parts per million, or one-tenth the concentration of previous additives (Fox 2006). Fuel economy benefits of up to 10% have been demonstrated in independently assessed field trials under commercial operating conditions. Additional pilot tests involving a range of vehicles and engines are underway (Oxonica 2006).

More energy-efficient transportation resulting from the use of high-strength, low-weight materials developed with nanotechnology may reduce the need for transportation fuels that would be shipped via pipeline along a ROW. Nanoparticle-reinforced materials that are as strong as or stronger than today's materials but weigh less will help provide better fuel economy. By using high-strength nanomaterials, parts for automobiles and other modes of transportation could be more than 50% lighter than conventional alternatives.

The reduction in weight could cut fuel requirements, thereby potentially reducing the demand for 7 petroleum fuel (and its attendant pipeline transportation in ROWs). Similarly, new materials developed through nanotechnology will permit the miniaturization of systems and equipment, which may further improve fuel economy.

More efficient batteries developed by using better electrolytes (composed of nanomaterials) may also reduce the need for transportation fuels. Nanotechnology is being used in lithium-ion and other batteries that are expected to increase the efficiencies of hybrid and electric vehicles.

Nanoscale capacitors made from multiwalled CNTs dramatically boost the amount of surface area, and thus the electrical charge, that each metal electrode in the capacitor can possess. Smaller and more powerful capacitors may facilitate the development of microchips having greatly increased circuit density. Such nanoscale capacitors may also impact the development of compact and cost-effective supercapacitors, which could help reduce the amount of weight in hybrid-electric vehicles, thus improving fuel consumption (UPI 2006).

Nanotechnologies may also facilitate the generation of electricity directly from solar, wind, and geothermal sources. Using such energy at or near the source could enable distributed energy production of electricity, thereby minimizing transmission losses and reducing the need for ROW-based transmission of electricity, oil, and gas. Practical energy collectors that are simple and automated may result from cheap nanofabrication (Gillett 2002). Also, more efficient electricity transmission may enable the generation of increased amounts of electricity in remote locations (e.g., nonpopulated areas with abundant renewable energy) to be sent to high-energydemand areas via nanoenhanced transmission.

Solar photovoltaic technology, which at present relies on crystalline-silicon wafers that are costly to produce, is deployed economically only in limited settings. Less costly quantum dot (nanocrystal) technologies could make important contributions to improving the efficiency of solar energy systems.

High-performance semiconductor nanocrystals (nanodots) that are active over the entire visible spectrum and into the near-infrared have been combined with conductive polymers to create ultrahigh-performance solar cells. The solar cells have improved efficiencies because the nanocrystals harvest a greater portion of the energy spectrum. Solar roofing tiles using quantum dots that are based on metal nanoparticles are expected to be commercialized within the next several years (Strem 2006).

Highly ordered nanotube arrays have demonstrated remarkable properties when used in solar cells. Researchers explain that the nanotube arrays provide excellent pathways for electron percolation, acting as "electron highways" for directing the photo-generated electrons to locations where they can do useful 8 work. Research results suggest that highly efficient solar cells could be made simply by increasing the length of the nanotube arrays (Penn State 2006).

One solar energy company (Konarka Technologies) creates a photoactive nanoscale material that can be printed on a variety of surfaces, including flexible plastics that can be manufactured in rolls. The material can be cut up and used for such applications as roofing and interior wall material, and can even be stitched onto or woven into a soldier's backpack.

The product's costs are one-third those of conventional photovoltaics, and the projected capital cost for manufacturing equipment and facilities is about one-fifth that of the prevailing cost for conventional solar cells (McGahn 2006).

Nanoadditives, including nanoparticles and nanopowders, could be used to enhance the transfer of heat from solar collectors to storage tanks. When added to heat-transfer fluids, the solid nanoparticles conduct heat better than the fluids alone, and they stay suspended longer than larger particles (Strem 2006).

Energy Storage The ability to store energy locally can reduce the amount of electricity that needs to be transmitted over power lines to meet peak demands. Energy storage could allow downsizing of baseload capacity and is a prerequisite for increasing the penetration of renewable and distributed generation technologies such as wind turbines at reasonable economic and environmental costs. Suitable energy storage is critical to the increased use of renewable energies, particularly solar and wind, because these are inconsistent resources.

Nanotechnology may play a role in distributed generation through the development of cost-effective energy storage in batteries, capacitors, and fuel cells. The next generation of storage devices may be optimized by nanoengineered advances and the use of nanoscale catalyst particles (Foster 2006).

Richard Smalley has described a model for storage using nanotechnology (Foster 2006) in which he suggests that by 2050 every house, business, and building would have its own local electrical storage device – an uninterruptible power supply capable of handling all of the needs of the owner for 24 hours.

Because such devices would be small and relatively inexpensive, they could be replaced with new models every

5 years or so as technological innovation continues. Today, such a unit, using lead-acid storage batteries and storing 100-kilowatt-hours (kWh) of electrical energy for a typical house, would occupy a small room and cost more than $10,000.

Through advances in nanotechnology, it may be possible to shrink an equivalent unit to the size of a washing machine and drop the cost to less than $1,000. With these advances, the electrical grid could become exceedingly robust. Such advances could also permit some or all of the primary electrical power on the grid to come from solar and wind energy.

CNTs have extraordinarily high surface areas and good electrical conductivity, and their linear geometry makes their surface areas highly accessible to a battery's electrolyte. These properties could enable CNT-based electrodes in batteries to generate increased electricity output as compared to traditional electrodes. This ability to increase the energy output from a given amount of material means not only that batteries could become more powerful, but also that smaller and lighter batteries could be developed for a wider range of applications. Commercial firms are actually developing such next-generation batteries today.

For example, in April 2006, a nanotechnology company (Altairnano) owner testified before Congress about a new nanotechnology battery with potentially broad applications. The battery technology utilizes 25-nm nanostructured lithium titanate spinel (a hard, glassy mineral) as the electrode material in the anode of a rechargeable lithium-ion battery, replacing the graphite electrode typically used in such batteries and contributing to performance and safety issues.

The new battery offers vastly faster discharge and charge rates, meaning that the time to recharge the battery can be measured in minutes rather than in hours. The nanostructured materials also increase the useful lifetime of the battery by 10 to 20 times over current lithium batteries and provide battery performance over a broader range of temperatures than currently achievable; over 75% of normal power would be available at temperatures between −40°F and +152°F.

These types of batteries may enable the U.S. auto industry to "leapfrog" the next generation of hybrid-electric vehicles, thereby accelerating the reduction of the need for petroleum (and for the pipeline transmission of petroleum). Other commercial applications for these batteries are for uninterruptible power supplies (UPSs) and emergency backup power (EBP). Present-day UPS and EBP systems typically use lead-acid batteries because of their reliability and low initial cost. However, lead-acid batteries must be replaced every 2 to 3 years, and hazardous materials issues surround their manufacture, handling, and maintenance. Lead-acid batteries also lose charge quickly in extreme temperatures (112°F) and suffer from power declines and a decreased ability to accept a recharge over time. By comparison, prototype batteries using the nanostructured lithium titanate electrode material show promising improvements.

The advanced lithium-ion battery is virtually unaffected by temperature extremes, its charge is fully available immediately, and it can accept a full recharge in a few minutes. It also has a much longer lifetime with no decline in performance, and there are no hazardous materials issues. With these kinds of advantages, UPS and EBP systems could become reliable components of distributed mini-grids (Gotcher 2006).

While batteries, which derive electrical energy from chemical reactions, are effective in storing large amounts of energy, they must be discarded after many charges and discharges. Capacitors, however, store electricity between a pair of metal electrodes. They charge faster and longer than normal batteries, but because their storage capacity is proportional to the surface area of their electrodes, even today's most powerful capacitors hold 25 times less energy than similarly sized chemical batteries.

Researchers, however, have covered capacitor electrodes with millions of nanotubes to increase electrode surface area and thus the amount of energy that they can hold. The researchers claim that the new technology "combines the strength of today's batteries with the longevity and speed of capacitors and has broad practical possibilities, affecting any device that requires a battery" (Limjoco 2006).

Because conventional lithium-ion batteries cannot charge at temperatures below 32°F and explode at temperatures higher than 208°F, this characteristic would permit the new batteries to be used in physical environments that today cannot be served by lithium-ion batteries because of safety concerns or because they require complex, expensive electronic control circuitry and temperature maintenance.

A fuel cell is a device used for electricity generation that is composed of electrodes that convert the energy of a chemical reaction directly into electrical energy, heat, and water. It is similar to a battery, except that it is designed for continuous replenishment of the reactants that become consumed, thereby requiring no recharging. It produces electricity from an external supply of fuel and oxygen, rather than the limited internal energy storage capacity of the battery.

Fuel cells come in various sizes and provide useful power in remote locations such as spacecraft and weather stations. Fuel cells are often considered in the context of hydrogen, because they change hydrogen and oxygen into water, producing electricity and heat in the process but no other by-products. A fuel-cell system running on hydrogen has no major moving parts and can be compact and lightweight.

Many believe that in the future, fuel cells will be used to power everything from handheld electronic devices to cars, buildings, and utility power plants. IBM projects that fuel cells in cars will be a "daily fact of life" by 2010, and General Motors estimates that it will have a million fuel-cell cars in production by then (IBM 2004).

Such technologies may supply much of the power that would otherwise need to be transported via ROWs (although pipelines may still be needed to transport the natural gas or hydrogen that feeds the fuel cells). Fuel cells are not new, but the materials' costs and complex manufacturing processes have limited their development. Nanoengineered materials may help improve fuel cells' efficiency in several ways.

Fuel cells operate by catalyzing the conversion of hydrogen into energy as the hydrogen passes through a catalytic medium. Advanced designs for next generation fuel cells involve the use of a polymer membrane as the structure through which the hydrogen passes and on which the catalysis occurs. The use of nanoengineered membrane materials may increase the volume of hydrogen conversion and thus result in more energy (McGahn 2006).

Precious metal nanoparticles of various compositions have been optimized to act as effective electro-catalysts in polymer electrolyte fuel cells and direct methanol fuel cells at both the anode and the cathode sides (Strem 2006).

A materials design concept used to control and manipulate the structure of a new material on the nanoscale could lead to more powerful fuel cells than currently available and to devices that enable more efficient energy extraction from fossil fuels and carbon-neutral fuels. The new electrode material allows more efficient direct utilization of natural gas or biogas (produced from waste) in fuel cells (Ruiz-Morales et al. 2006).

CNTs' high strength and toughness-to-weight characteristics may be important for composite components in fuel cells that are deployed in transport applications where durability is important.

Hydrogen Generation and Storage The hydrogen economy is a hypothetical future economy in which hydrogen is the primary form of stored energy for mobile applications and load balancing. It is typically discussed as an alternative to today's fossil-fuel economy. Many barriers need to be overcome for the hydrogen economy to become a reality.

These include producing the hydrogen (for which adequate sources of electricity are needed), transporting it (including the possible need for additional ROWs), and storing it. Nanotechnology may play a role in helping to meet these challenges. Nanotechnology may help accelerate the use of solar and other renewables to generate electricity.

Because hydrogen is the smallest element, it can escape from tanks and pipes more easily than conventional fuels. There are two ways to store hydrogen in materials. One way involves absorption of the hydrogen within the material, and the other is to store the hydrogen in a container. The challenge for absorption is to control the diameter of the nanotube so that the absorption energy of hydrogen on the outside and inside If successful, these small solid-state devices could improve current power generation and waste heat recovery techniques.

They are estimated to deliver up to 70 to 80% of the maximum (Carnot) theoretical efficiency for heat pumps (conventional power-generation equipment operates at up to 40% Carnot efficiency). Currently under development, Power Chips contain no moving parts or motors and can be either miniaturized or scaled to very large sizes for use in a variety of applications.

Numerous nanomaterials and other nano-related applications relevant to electricity transmission and petroleum distillate fuel and gas pipeline transport are in various stages of research, development, and deployment. These applications have the potential to directly or indirectly reduce the environmental impact associated with the construction, operation, and dismantlement of energy transmission technologies.

Potential pitfalls and timeframes have been identified in the literature. In general, however, the potential for practical scale-up of most of the techniques in use today for nanoparticle production is limited by high capital costs, low production rates, the need for exotic and expensive precursor materials, and limited control over nanoparticle physical and chemical homogeneity. Breakthroughs in nanotechnology research may accelerate the development and implementation of these technologies.

Wires and Cables Nanotechnology may help improve the efficiency of electricity transmission wires. Today, aluminum conductor steel reinforced (ACSR) wire is the standard overhead conductor against which alternatives are compared. By 2010, the development of new overhead conductors is expected to increase the capacity of existing ROWs by five times that of ACSR wire at current costs (DOE 2006).

The 3M Corporation has developed a nanomaterial-based metal-matrix overhead conductor known as the aluminum conductor composite reinforced (ACCR) wire, which is designed to resist heat sag and provide more than twice the transmission capacity of conventional conductors of similar size.

This ACCR wire is currently in use, or has been selected for use, by six major utilities across the country. According to 3M (2006): "Aluminum has been a key ingredient in bare overhead conductors for decades. The difference is that ACCR wire is based on the use of aluminum processed in new ways to create high-performance and reliable overhead conductors that retain strength at high temperatures and are not adversely affected by environmental conditions."

The ACCR wire's strength and durability derive from its nanocrystalline aluminum oxide fibers, which are embedded in the high-purity 3M aluminum matrix core wires using a patented process. The constituent materials are chemically inert with respect to each other and can withstand extreme temperatures without chemical reactions or any appreciable loss in strength.

The material used in the core of the cable replaces the steel used in conventional cables (3M 2006). Another example, mentioned in Section 2.2 and still in the research phase, is the use of armchair CNTs − a special kind of single-walled CNT that exhibits extremely high electrical conductivity (more than 10 times greater than copper).

Also possessing flexibility, elasticity, and tremendous tensile strength, CNTs have the potential, when woven into wires and cables, to provide electricity transmission lines with substantially improved performance over current power lines (Technology Administration 2003).

Replacing current wires with nanoscale transmission wires, called quantum wires (QWs) or armchair QWs, could revolutionize the electrical grid. The electrical conductivity of QW is higher than that of copper at one-sixth the weight, and QW is twice as strong as steel. A grid made up of such transmission wires would have no line losses or resistance, because the electrons would be forced lengthwise through the tube and could not escape out at other angles.

Grid properties would be resistant to temperature changes and would have minimal or no sag. (Reduced sag would allow towers to be placed farther apart, reducing footprint and attendant construction and maintenance impacts.) QW, if spun into noncorrosive polypropylene-like rope, could conceivably be buried "forever" with no fear of corrosion and "no need for shielding of any kind" (Hoffert 2004).

Such a grid could have a million times greater capacity than what exists today (assuming the 1-centimeter-diameter aluminum cable carrying about 1,000 to 2,000 amps); even if the capacity were increased by only 0.1%, the amount of enhanced capacity would still be impressive (Davis 2006). The realization of such conducting possibilities depends on developing processes for producing high-quality CNTs in industrial quantities and at reasonable cost, finding ways to manipulate and orient nanotubes into regular arrays, and developing robust testing methods.

Today, QWs made from metallic CNTs are very short – no longer than several centimeters – and are manufactured only in limited quantities. When nanotubes are synthesized, a variety of different configurations appear. (The armchair CNT is the only type that conducts electricity well enough for QWs.) Currently, only 2% of all nanotubes can be used as QWs, and sorting the armchair nanotubes from the rest is nearly impossible.

Current processing technologies are not capable of producing nanotubes with controlled and desirable production properties consistently. Until a good solution for separating the "good one from the many other unfavorable configurations is reached for large-volume manufacturing, the impact of nanotubes on power line usage is hypothetical" (Davis 2006).

The National Aeronautics and Space Administration has funded research to produce a 1-meter (m)-long prototype of QW by 2010. It is estimated that at least 5 years will be needed to develop methods to produce QW with high enough purity, in large enough amounts, and cheaply enough to spin continuous fibers into QW (Anderson et al. 2006). CNT manufacturers and government officials express optimism about the successful deployment of CNT in transmission wires, partly because, with a rope of CNTs woven together, it is not necessary for any single fiber to span the entire length of a transmission wire, since quantum tunneling allows electrons to jump from stand to strand.

The president of Raymor Industries (a company specializing in the development and application of CNTs) said in 2006 that "a transmission wire product will be commercially available in the 'not too distant future' but as of yet, no single-walled carbon nanotube provider has been able to demonstrate their ability to supply the material in large volumes with reasonable pricing, which is the only path to adaptation of this technology across the power grid" (Davis 2006).

Ataf Carim, with the U.S. Department of Energy, said earlier, "While promoting carbon nanotubes to electric utility companies in the near term would be premature, we certainly hope that the promise of such materials for power transmission can be realized in the future" (Davis 2006). Long-distance transmission of electrical current entails significant losses (about 20%) due to electrical resistance.

Superconductors transmit electricity with a small fraction of the losses from conventional conductors, thereby enabling power transmission at higher power densities. Such efficiencies may relieve transmission congestion and lessen the need for transmission equipment. High-temperature superconductors (i.e., substances that become superconducting near liquid nitrogen temperatures [about 77 Kelvin (K)] rather than near liquid helium temperatures [about 4 K]) were discovered in the late 1980s.

Noting that transmission constraints have contributed to higher electricity prices and reduced reliability, the 2001 National Energy Policy Report (National Energy Policy Development Group 2001) recommended expanded research and development on transmission reliability and superconductivity. HTS cables have been demonstrated at full scale at distribution voltages and in lengths up to 100 m. Large-scale use of second-generation HTS wire carrying high-amperage electrical current with virtually no resistance promises dramatic gains in energy efficiency. Other advantages of HTS cables include: HTS cables can carry more power at the same voltage than conventional cables, meaning that the need for 500-kilovolt (kV) and higher voltage transmission, which requires expensive power equipment, could be eliminated.

Because HTS cables are operated at cryogenic temperatures, they have a lower susceptibility to temperature-related faults than overhead lines.

HTS cables would be sited underground, making them less vulnerable than overhead transmission lines to natural events and unintentional or intentional disruptions. HTS also has disadvantages. For one, the cost of the HTS wire is currently much higher than that of conventional conductors (such as copper or aluminum) used for electricity transmission and distribution.

Also, HTS wires are more susceptible to magnetic fields than their metallic counterparts. If exposed to large magnetic fields such as those found in transmission lines, they would develop resistance, which would heat the ceramic and result in decreased efficiency.

Nanotechnology may be helpful in mitigating some of these challenges, bringing HTS closer to commercial use. For example, American Superconductor Corporation has developed and filed a patent application for a nanotechnology-based manufacturing technique that delivers an immediate 30% increase in the electric current-carrying capability of the company's second generation (2G) HTS wire.

This process disperses nanodots throughout the superconductor coating of the company's 2G HTS wire. The nanodots are ultrasmall particles (typically less than 100 atoms across) of inorganic materials that increase the flow of electrical current through the 2G HTS wire by pinning (immobilizing) magnetic lines of flux in the superconductor (Nanotechwire 2004a).

The pinning allows higher amounts of electrical current to flow even in the presence of strong magnetic fields and at relatively high operating temperatures. The 2G HTS wire is being designed as a replacement for today's commercial first-generation (1G).

HTS wire at manufacturing costs that are two to five times lower than those of the 1G HTS wire. Nanotechnology applications may help improve other components of the electric transmission infrastructure, thereby potentially reducing environmental impacts. The examples below pertain to transformers, substations, and sensors.

Fluids containing nanomaterials could provide more efficient coolants in transformers, possibly reducing the footprints, or even the number, of transformers. Nanoparticles increase heat transfer, and solid nanoparticles conduct heat better than liquid. Nanoparticles stay suspended in liquids longer than larger particles, and they have a much greater surface area, which is where heat transfer takes place (Strem 2006).

Using nanoparticles in the development of HTS transformers could result in compact units with no flammable liquids, which could help increase siting flexibility.

Substation batteries are important for load-leveling peak shaving, providing uninterruptible supplies of electricity to power substation switchgear, and for starting backup power systems. Smaller, more efficient batteries could reduce the footprints of substations and possibly the number of substations within a ROW.

Nanoelectronics have the potential to revolutionize sensors and power-control devices. Nanotechnology-enabled sensors would be self-calibrating and self-diagnosing. They could place trouble calls to technicians whenever problems were predicted or encountered. Such sensors could also allow for the remote monitoring of infrastructure on a real-time basis.

Miniature sensors deployed throughout an entire transmission network could provide access to data and information previously unavailable. The real-time energized status of distribution feeders would speed outage restoration, and phase balancing and line loss would be easier to manage, helping to improve the overall operation of the distribution feeder network.

Other Materials Advanced materials using nanomaterials or nanotechnology may extend service life, lower failure rates, and reduce the potential for environmental damage. Two examples of such advanced materials follow.

The Electric Power Research Institute, Inc. (EPRI) has described how nanotechnology may accelerate the development of "smart materials and structures" (SMSs) (EPRI 2003). According to EPRI, SMSs have the unique capability to sense and physically respond to changes in their environment (e.g., temperature, acidity levels, and magnetic field). Generally consisting of a sensor, an actuator, and a processor, an SMS device can function autonomously in an almost biological manner.

On a transmission line, SMSs could monitor and assess the condition of conductors, breakers, and transformers in real time to avoid outages. SMSs may also enable in situ repair of underground cables. In addition, smart materials may be used to adjust transmission line loads according to real-time thermal measurements. However, to realize such capabilities, more research and development are required to integrate SMSs into components, embed the SMS components into the structure to be controlled, and facilitate communication between smart structure components and the external world. EPRI also notes that nanoscale electronics will most likely be used to make circuits smaller than they are now.

Such circuits would be molecular-scale, high-speed, high-capacity electronic circuits. EPRI notes that while basic transistors have been created from organic molecules, the feasibility of building complicated nanoscale electronic or mechanical devices will require solutions to fabrication problems, quantum effects problems, and communication difficulties (EPRI 2003).

Intelligent substations that redirect electrical flow around congestion and take actions determined by simulations will be aided by new power-controller hardware. Smart power controllers employ computer control of the electricity grid's equivalent of transistors and giant capacitors to divert power from troubling congestion areas to underutilized grid lines.

Computers operating these power controllers at every level of the grid would reduce the need to build new generation through better efficiency and safety. However, power-controller hardware is big, heavy, and expensive.

Nanomaterials may provide a new generation of power controllers that are cheaper, stronger, and lighter than today's prototypes (Anderson et al. 2006).

Ceramics are hard and resist heat and chemical attack, but they are also very brittle. Researchers at the University of California at Davis have mixed aluminum oxide with 5% to 10% CNTs and 5% finely milled niobium and processed it to consolidate ceramic powders at lower temperatures than conventional processes. The resulting material has up to five times the fracture toughness (resistance to cracking under stress) of conventional alumina.

The material also shows electrical conductivity seven times that of previous ceramics made with nanotubes. It also has interesting thermal properties – conducting heat in one direction, along the alignment of the nanotubes, but reflecting heat at right angles to the nanotubes – making it an attractive material for thermal barrier coatings (Foley 2003).

Today, most of the identified nanotechnology applications for pipelines involve material coatings (insulation, corrosion, and multipurpose). Other potential applications include nanosensors, which have the potential to minimize environmental damage by identifying potential leaks before they spread, and oil spill remediation with nanomaterials, which may minimize damage should a leak occur. Because the current and expected future applications of nanotechnology for petroleum distillate fuel pipelines are basically the same as those for natural gas pipelines, this section cites examples of general nanotechnology applications for pipelines.

Advanced materials using nanotechnology may extend service life, reduce failure rates, and limit the potential for environmental damage. Nanocoating metallic surfaces can help achieve superhardening, low friction, and enhanced corrosion protection. Stronger materials may reduce wear, corrosion, and the chances of puncturing associated with third-party damage.

Also, because nanomaterials can be stronger per unit volume than conventional materials, the use of pipe materials that contain or are coated with nanomaterials may mean fewer disturbances to the environment during installation, maintenance, and dismantlement.

Corrosion under insulation (CUI) is a costly problem that is difficult to detect in pipelines. According to a study by CC Technologies in cooperation with NACE International, the costs associated with direct corrosion on gas and liquid transmission pipelines are $7 billion per year. Corrosion was the major cause of reportable incidents in North America requiring more than $1 billion in repairs to one pipeline alone.

Nansulate™ is a high-performance thermal insulator that prevents CUI. The translucent characteristic of the coating allows for visual inspection of the substrate without having to remove the coating, making it well-suited for use in gas and liquid transmission pipelines. The coatings utilize nanotechnology to prevent CUI, whereas many of the insulations currently in use actually cause the problem of CUI (Industrial Nanotech, Inc. 2006).

Nanostructure coatings have excellent toughness, wear, and adhesion properties. Nanostructure powders have grains less than 100 nm in size, which are agglomerated to form particles large enough to be sprayed using conventional thermal spray methods. These coatings may be used to repair component parts instead of replacing them, resulting in reduced maintenance costs and disturbance. Additionally, the nanostructure coatings may extend the service life of the components because of their improved properties over conventional coatings (Vasanth and Taylor 2002).

Aerogels are highly porous solids formed from a gel in which the liquid is replaced with a gas. Containing more than 95% air, they have been dubbed the world's lightest solids. While aerogels were first discovered in 1931, their insulating power was unusable because they were so brittle.

In 2001, Aspen Aerogels commercialized a process to deliver aerogels in flexible blankets, making aerogels easy to use almost anywhere. Using nanotechnology, Aspen now produces the same insulating material in a flexible foam infused with silica nanostructures. Because of the formulation, less insulation can be used to accomplish the same objectives, resulting in smaller pipe diameters (smaller footprint). Also, because heat radiation is a function of surface area, up to 30% less heat is dissipated (Aspen 2006).

A new class of material known as Quasam™ can be produced as either a thin-film coating or as a bulk material. The material exhibits ultra-low weight combined with the strength, stiffness, hardness, and thermal stability approaching that of natural diamond. The material demonstrates chemical and corrosion resistance, biocompatibility, and electrical conductivity even at a 1-nanometer thickness.

It also has the ability to act as a "smart skin," which provides the ability to measure stress and detect structural damage before an incident. A Quasam coating on an oil pipeline could provide detection of stress buildup before a failure occurs (Nanotechwire 2004b).

Nanosensors, or sensors made of nanomaterials, can be extremely sensitive, selective, and responsive. As such, they could be smaller and cheaper, and consume less power than conventional sensors. Sensors and controls that are small in size; work safely in the presence of electromagnetic fields, high temperatures, and high pressures; and can be changed cost-effectively may provide the ability to monitor conditions in the infrastructure and monitor for pollutants (vapor or oil losses) continually.

Researchers have recently developed sensors made from titania nanotubes coated with a discontinuous layer of palladium. The photocatalytic properties of titania nanotubes are so large – a factor of 100 times greater than any other form of titania – that sensor contaminants are efficiently removed with exposure to ultraviolet light, so that the sensors effectively recover or retain their original sensitivity to hydrogen. Sensors in uncontrolled locations become contaminated by a variety of substances including volatile organic vapors, carbon soot, and oil vapors, as well as dust and pollen.

A selfcleaning function capable of oxidizing contaminants would extend sensor lifetime and minimize sensor errors (Science Daily 2004). Nanosensors may also be used to identify approaching vehicles or equipment that may otherwise lead to third-party damage; or, if the pipeline is damaged, provide an immediate indication (e.g., alarm or other notification) so that any potential environmental damage may be mitigated quickly.

A substance that uses nanoparticles and a patented "selfassembled monolayer" (SAM) technology has been developed to absorb about 40 times its weight in oil. Because water is completely rejected by the material, spilled oil can be recovered for use – a substantial benefit in oil spill cleanup efforts. The substance combines materials and surface innovation at the nanoscale level with SAM technology to produce a class of nanostructured hybrid materials that serve as environmental sorbent materials.

The system provides the ability to form chemical foundations from which other building blocks can be used to form more complex structures. The interfacial chemistry can be engaged with specific polymer systems to enhance the adhesion properties of a polymer to a solid substrate. 19 The technology may provide advancements in oil spill recovery and remediation (Lamba 2005).

Advances such as this may offer the potential for reduced environmental damage to ROWs and nearby areas, should a leak occur.

Due to their extremely small size and relatively large surface areas, nanomaterials may interact with the environment in ways that differ from more conventional materials. Potentially harmful effects of nanotechnology could result from the nature of the nanoparticles themselves and from products made with them. Environmental, safety, and health (ES&H) risks could occur during research, development, production, use, and end-of-life processes.

The 21st Century Nanotechnology Research and Development Act, signed into law in December 2003, calls for addressing potential environmental and societal concerns associated with nanotechnology. About 10% of the Federal nanotechnology budget is characterized as environmental, but much of this amount is for developing nanotechnologies to address existing environmental problems rather than for investigating potential ES&H effects.

Some research has been conducted on the toxicology of nanomaterials, but research on the fate, transport, and transformation; risk characterization, mitigation, and communication; and exposure, bioaccumulation, and personal protection have yet to come. All of the departments and agencies with Federal funding from the National Nanotechnology Initiative have environmental research planned or underway, but industry, nongovernmental organizations, and others have questioned whether the amount of research on the potential ES&H impacts of nanotechnologies is sufficient, given the number of unknowns and the size of the potential nanotechnology market.

There have been relatively few studies on the ES&H risks of engineered nanoparticles. There are also no regulatory requirements to conduct such studies, and little funding is allocated to them. The limited results to date are neither conclusive nor consistent. For example, evidence indicates that nanoparticles in the lungs may cause more severe damage than conventional toxic dusts, but few if any inhalation or exposure studies have been conducted. Nanotechnologies may speed cleanup of soil and water contamination, but in the process may harm local soil ecology. The impacts of large quantities of nanomaterials on the environment or human health have not been studied, and there are no studies on accumulation or other long-term impacts.

5. THE USES AND APPLICATIONS OF NANOTECHNOLOGY IN MEDICE

Nanotechnology is the term derived from the Greek word "nano" meaning "dwarf" (short man). Nanomedicine involves utilization of nanotechnology for the benefit of human health and well-being. A nanometer is a billionth of a meter. It is difficult to imagine anything so small, but think of something only 1/80,000 the width of a human hair. Ten hydrogen atoms could be laid side-by side in a single nanometer. A red blood cell is approximately 7000 nm wide. Many molecules including some proteins range between 1 nm and larger. Nanotechnology is the creation of useful materials, devices, and systems through the manipulation of matter on this minute scale.

The emerging field of nanotechnology involves scientists from many different disciplines, including physicists, chemists, engineers, and biologists. Although the word nanotechnology is relatively new, the "natural version" of nanotechnology was already in pole position with procreation of life itself thousands of millions of years ago. All natural materials and systems establish their foundation at the nanoscale. Basically, the biological building blocks of life are nano-entities that possess unique properties determined by the size, folding and patterns at nanoscale. The genetic material deoxyribonucleic acid (DNA) is composed of four nucleotide bases in sizes ranging in the sub-nanometre scale, and the diameter of the double helix structure of DNA is in the nanometre range. The same is true for proteins and cell membranes which consist of lipids and proteins.

The nanometer scale of nanomedicines is considered to be ideal to interact with cells that have dimensions (microns) that allow them to efficiently interact with nanoparticles (10–200 nm). Many drugs can be made significantly more effective if delivered using appropriate drug-delivery vehicles that allow them to efficiently reach their target in a form that both enables the drugs to be taken up by cells and minimizes off-target effects. More efficient and accurate delivery to their targets is expected to reduce the side effects of nanomedicines compared to conventional delivery.

Understanding of biological processes on the nanoscale level is a strong driving force behind development of nanotechnology. In essence, Nanomedicine is the medical use of nano-sized particles, nanofiber and nanodevices to deliver drugs, heat, light or other substances to specific cells in the human body and for the detection and treatment of diseases or injuries within the targeted cells, thereby minimizing the damage to healthy cells in the body.

Different materials interact differently with their environment when they are sized on the nanoscale. Bulk materials interact with their environment in a certain way because the vast majority of their atoms are inside the volume of the material rather than on the surface, this makes the surface to-volume ratio very small .Atoms respond to their environment differently when they are surrounded by other atoms than when they are on a surface and do not have atoms surrounding them. And the relative amount of atoms on the surface can greatly influence the properties of the material as a whole. With nanoscale materials, many of the atoms reside on the surface of the material and therefore the surface-to-volume ratio is much larger.

From the biological point of view, nanodevices match the typical size of naturally occurring functional units or components of living organisms and, for this reason, enable more effective interaction with biological systems. Most animal cells are 10,000 to 20,000 nanometers in diameter. This means that nanoscale devices (less than 100 nanometers) can enter cells and the organelles inside them to interact with DNA and proteins. Nanodevices developed through nanotechnology are able to detect disease in a very small amount of cells or tissue. They can enter and monitor cells within a living body. Fig 2. Shows a nano device inside an animal cell. Because of their small size and larger surface area relative to their volume, nanoscale devices can readily interact with bio molecules such as enzymes and receptors on both, the surface of the cell and inside the cell. By gaining access to various areas of the body, nanoparticles have the potential to detect diseases at the micro level and deliver treatment. It is now possible to provide therapy at a molecular level with the help of nanotechnology. At the nanoscale the characteristics of matter can be significantly changed, particularly under 10- 20 nm, because of properties such as the dominance of quantum effects, confinement effects, molecular recognition and an increase in relative surface area [6]. Downsized material structures of the same chemical elements change their mechanical, optical, magnetic and electronic properties, as well as their chemical reactivity, leading to surprising and unpredictable effects. In short, nanodevices exist in a unique kingdom, where the properties of matter are governed by a complex combination of classical physics and quantum mechanics.

In order to interact with biological target, a biological or molecular coating or layer acting as a bioinorganic interface should be attached to the nanoparticle. Examples of biological coatings may include antibodies, biopolymers like collagen, or monolayers of small molecules that make the nanoparticles biocompatible.

In addition, as optical detection techniques are wide spread in biological research, nanoparticles should either fluoresce or change their optical properties. Nano-particle usually forms the core of nanobiomaterial. It can be used as a convenient surface for molecular assembly, and may be composed of inorganic or polymeric materials. It can also be in the form of nanovesicle surrounded by a membrane or a layer. The shape is more often spherical but cylindrical, plate-like and other shapes are possible.

The core particle is often protected by several monolayers of inert material, for example silica. The same layer might act as a biocompatible material. However, more often an additional layer of linker molecules is required to proceed with further functionalisation. This linear linker molecule has reactive groups at both ends. One group is aimed at attaching the linker to the nanoparticle surface and the other is used to bind various moieties like biocompatibles (dextran), antibodies, fluorophores etc., depending on the function required by the application.

Nanoparticles can be prepared either by the "top down" technique starting with large particles and making things smaller by grinding or pulverizing. This approach has traditionally been used in making parts for computers and electronics. The other technique is "bottom up" technique. This helps making things larger by building atom by atom or molecule by molecule and may prove useful in manufacturing devices used in medicine.

The control of the process is critical for the production of nanostructures. The development of enhanced microscopy techniques such as scanning tunneling microscopy (STM) and atomic force microscopy (AFM), has facilitated the use of the bottom up process. Most of the time nanoparticles are fixed with or on the surface of material. Nanoparticles can be made from a vast range of materials, such as metals (gold/silver), metal oxides (e.g. titanium dioxide -TiO_2, Silicon dioxide SiO_2), Inorganic materials (carbon nanotubes, quantum dots), polymeric materials and lipids.

The other new sets of tools is available in nanotechnology are nanocrystals (Quantum dots), cantilevers, dendrimers, nanoshells and nanowires. These particles range from few to several hundreds of nanometers in diameter. Products made from each of these tools can be used for diagnosis (as biomarkers) and therapy.

Nanotechnology in healthcare management includes Liposomes, Quantum dots (Nanocrystals), Nanoshells, Cantilevers- A Nanoelectromechanical Sensor (NEMS), Dendrimers and Nanowires.

Liposomes were discovered in early 1960"s by Bangham and collegue. When dry phospholipids, or a mixture of such phospholipids and cholesterol, are immersed in water under laboratory conditions, they spontaneously form closed structure with internal aqueous environment bounded by phospholipid bilayer membranes known as liposome. Liposomes are small vesicle (a fluid filled sac) of spherical shape.

They are biodegradable, biocompatible and non-immunogenic in nature, which makes them ideal drug carrier systems in therapeutics. Liposomes are made of concentric spheres, one sphere inside of another and each forming half of a bilayered wall. A bilayer is composed of two sheets of phospholipid. Liposomes are useful for encapsulating other molecules such as pharmaceutical drugs.

Liposomes are drug carrier loaded with great variety of molecules such as small drug molecules, proteins, nucleotides and even plasmids. Liposomes are lipid based nanoparticles used extensively in pharmaceutical and cosmetic industries because of their capacity for breaking down inside cells, once their delivery function has been met. Liposomes were first engineered nanoparticles used for drug delivery.

Cancer chemotherapeutic drugs and other toxic drugs like amphotericin and hamycin, when used as liposomal drugs produce much better efficacy and safety as compared to conventional preparations. These liposomes can be loaded with drugs either in the aqueous compartment or in the lipid membrane. Liposomes transport hydrophilic drugs (water soluble) inside the core (aqueous compartment) and hydrophobic drugs (water insoluble but soluble in lipid) between the bilayer.

Quantum dots (QD) are tiny (~2-10 nanometers or ~10-50 atoms in diameter) light emitting semiconductor/nanocrystals with vast applications in medical sciences. They obey quantum mechanical principle of quantum confinement. These can be used to detect cancer in the body. Quantum dots when used in conjunction with magnetic resonance imaging can produce exceptional images of tumor sites. Quantum dots glow when they are stimulated by ultraviolet light. The wavelength, or color, of the light depends on the size of the crystal.

Latex beads (polymeric particles suspended in latex) filled with these crystals can be designed to bind to specific DNA sequences. By combining different sized quantum dots within a single bead, scientists can create probes that release distinct colors and intensities of light. When the crystals are stimulated by UV light, each bead emits light that serves as a sort of spectral bar code, identifying a particular region of DNA.

Another advantage of quantum dots is that they can be used in the body, eliminating the need for biopsy. Quantum dots are 20 times brighter & 100 times more stable than traditional dyes.

Nanoshells-tiny spheres of glass coated with gold are the first engineered nanomaterial to enter into human trials. Metal nanoshells are excellent optical absorbers. Particularly gold, because of the strong optical absorption from the metal's response to light. Similar to quantum dots, nanoshell diameter/size and thickness play a vital role in optical tuning of certain wavelength.

To achieve more effective and better diagnostic and therapeutic goals, nanoshells can be conjugated to antibodies, oligonucleotides, fluorophores, targeting ligands, polymers, therapeutic agents, and radioisotope. Nanoshells (especially gold nanoshells) show promise application in biomedical imaging, target therapy, gene delivery, tissue welding, drug delivery systems, therapeutic applications in general and cancer imaging and treatment in particular.

For treatment, a cancer patient receives a dose of nanoshells intravenously, and over the course of a day about 1% accumulates in a tumor site. Most of the rest wash out being so small in size. A physician then shines an infrared light over the tumor. The absorption of light by the nanoshells creates an intense heat burning away the tumor, while healthy cells nearby remain unharmed. This killing effect of heat is also known as Hyperthermia.

Advantage: zero toxic effects (unlike chemotherapy) no ionizing radiation (like radiotherapy). Nanoshells loaded with insulin can be injected under the skin, where they can stay for months. To release the drug, patients use a pen-sized IR laser over the skin at the injection site.

The advances in micro- and nanofabrication technologies enable the preparation of increasingly smaller mechanical transducers capable of detecting the forces, motion, mechanical properties and masses that emerge in biomolecular interactions and fundamental biological processes. Thus, biosensors based on nanomechanical systems (NMS) have gained considerable relevance in the last decade. Due to the biological adsorption or interactions between the analyte (substance of interest, e.g. a particular chemical component, virus or micro-organism) and surface of cantilever some mechanical phenomena occur which shows a biological response.

Nanoscale cantilevers work on this principle. The surface of cantilever is coated with bioreceptor/biorecognition element. Nanoscale cantilevers - microscopic, flexible beams resembling a row of diving boards - are built using semiconductor lithographic techniques. A biosensor consists of two components, a bioreceptor and a transducer. The bioreceptor is a biomolecule that recognizes the target analyte thereby generating a comprehensive surface stress which causes a downwards bending of the cantilever whereas the transducer converts the recognition event into a measurable signal. They are also known as Nanoelectromechanical Systems (NEMS).

The physical properties of the cantilevers change as a result of the binding event. Researchers can read this change in real time and provide not only information about the presence and the absence but also the concentration of different molecular expressions. Nanoscale cantilevers, constructed as part of a larger diagnostic device, can provide rapid and sensitive detection of cancer-related molecules.

The word dendrimer originates from the Greek dendron, meaning 'tree' and meros meaning "part". A dendrimer consists of molecular chains that branch out from a common center (like a tree), and there is no entanglement between each dendrimer molecules. Biodendrimers comprised of repeating units known to be biocompatible or biodegradable in vivo to natural metabolites. Dendrimers are defined by three components; a central core, an interior dendritic structure (the branches), and an exterior surface with functional surface groups.

The interior of the dendrimer offer cavities (nanosized "container") that can readily accept small molecules or particles, which make them ideally suitable for encapsulation, isolation from external media or active catalytic sites. The surface of the dendrimer can be modified with a wide variety of function, allowing to finely tuning the chemical, physical and topological properties of the molecule.

The ends of the dendrimer molecule can be attached with other molecules for transport. These molecules give the dendrimers various functional applications. They are used in medical sciences for targeted drug delivery and contrast agent in MRI. The cavities present in dendrimers can be used as binding sites for smaller molecules - effectively the dendrite becomes a nanosized "container" for various molecules. A nanowire is a wire of diameter of the order of nm.

Typically their width ranges from forty to fifty nanometers, but their length is not so limited. Since they can be lengthened by simply attaching more wires end to end or just by growing them longer, they can be as long as desired. The nanowires have unique metallic, semiconducting, and insulating properties. The extremely high surface-to-volume ratio of 1D (1 dimension) biosensor like nanowires and nanotubes makes them ideal building blocks for biosensor development. They have strong biocompatibility and size similarity with the host (biomolecules), with emphasis on novel electron transport properties

. Nanowires can detect even slight disturbances from the surrounding environment (due to their high surface area to volume ratio). The constituent atoms reside on the surface of the nanostructures, which can generate electrical signals even with slight disturbances in the system. Nanowires are far smaller than the smallest capillary in the body that means nanowires could, in principle, be threaded through the circulatory system to any point in the body without blocking the normal flow of blood or interfering with the exchange of gases and nutrients through the blood-vessel walls.

Bunch of nanowires being guided through the circulatory system to the brain. Once there, the nanowires would spread out branching into tinier and tinier blood vessels. Each nanowire would then be used to record the electrical activity of a single nerve cell, or small groups of nerve cells (better than PET or fMRI) giving the ability to pinpoint damage from injury and stroke, localize the cause of seizures, and other brain abnormalities. PET-Scan stands for Positron emission tomography.

It is an imaging technique that helps to reveal how the tissues and organs are functioning in the body. It uses a radioactive drug/tracer to show this activity. The tracer may be injected, swallowed or inhaled depending on which organ or tissue is being studied by PET scan. The tracer collects in areas of the body that have higher levels of chemical activity, which often correspond to areas of disease. On a PET scan, these areas show up as bright spots. fMRI refers to as functional magnetic resonance imaging. It is a technique which is used to measure brain activity. It works by detecting the changes in blood oxygenation and flow that occur in response to neural activity – when a brain area is more active it consumes more oxygen and to meet this increased demand blood flow increases to the active area.

fMRI can be used to produce activation maps showing which parts of the brain are involved in a particular mental process. It's long been known that people with Parkinson's disease (a progressive disorder of the nervous system that affects the movement. It develops gradually, sometimes starting with a barely noticeable tremor in just one hand) can experience significant improvement from direct stimulation of the affected area of the brain with electrical pulses.

Indeed, that is now a common treatment for patients who do not respond to medication. But the stimulation is currently carried out by inserting wires through the skull and into the brain, a process that causes scarring of brain tissue. The hope is, by stimulating the brain with nanowires threaded through pre-existing blood vessels, doctors could give patients the benefits of the treatment without the damaging side effects. The small size of nanoparticles can be very useful in oncology, particularly in imaging.

In recent years significant efforts have been made to use nanotechnology for the purpose of drug and vaccine delivery. The nanoparticles offer a suitable means to deliver small molecular weight drugs as well as macromolecules such as proteins, peptides or genes in the body using various routes of administration. The nano-sized materials provide a mechanism for local or site specific targeted delivery of macromolecules to the tissue/organ of interest, in-vivo.

The newer developments in material science and nanoengineering are currently being leveraged to formulate therapeutic agents in biocompatible nanocomposites such as nanoparticles, nanocapsules, micellar systems and conjugates. This can be achieved by molecular targeting by nanoengineered devices.

It is all about targeting the molecules and delivering drugs with cell precision. Drug delivery systems, lipid- or polymer-based nanoparticles, can be designed to improve the pharmacological and therapeutic properties of drugs. The basic point to use drug delivery is based upon three facts, (a) Efficient encapsulation of the drugs, (b) Successful delivery of said drugs to the targeted region of the body and (c) Successful release of that drug there. Polymer-based nanoparticles are submicron-sized polymeric colloidal particles in which a therapeutic agent of interest can be embedded or encapsulated within their polymeric matrix or adsorbed or conjugated onto the surface.

These nanoparticles serve as an excellent vehicle for delivery of a number of biomolecules, drugs, genes and vaccines to the site of interest in-vivo. During the 1980's and 1990's several drug delivery systems were developed to improve the efficiency of drugs and minimize toxic side effects.

There is a size limit for the particles to be able to cross the intestinal mucosal barrier of the gastrointestinal (GI) tract after the drug has been delivered orally. Most often, macroparticles could not cross mucosal barrier due to their bigger sizes resulting in failed delivery of drugs. Nanoparticles on the other hand have an advantage over microparticles due their nano-sizes. Now, a wide variety of biomolecules, vaccines and drugs can be delivered into the body using nanoparticulate carriers and a number of routes of delivery.

NPs (nanoparticles) can be used to safely and reliably deliver hydrophilic drugs (having a strong affinity for water and readily dissolve in water), hydrophobic drugs (not to dissolve in water), proteins, vaccines, and other biological macromolecules in the body. They can be specifically designed for targeted drug delivery to the brain, arterial walls, lungs, tumor cells, liver, and spleen. They can also be designed for long-term systemic circulation within the body. In addition, nanoparticles tagged with imaging agents offer additional opportunities to exploit optical imaging or MRI in cancer diagnosis and guided hyperthermia therapy.

It can combine a specific targeting agent (usually with an antibody or peptide) with nanoparticles for imaging (such as quantum dots or magnetic nanoparticles), a cell-penetrating agent (e.g., the polyArg peptide TAT), a stimulus-selective element for drug release, a stabilizing polymer to ensure biocompatibility polyethylene glycol (most frequently), and the therapeutic compound. Development of novel strategies for controlled released of drugs will provide nanoparticles with the capability to deliver two or more therapeutic agents.

The small size of nanoparticles can be very useful in oncology, particularly in imaging. Quantum dots when used in conjunction with magnetic resonance imaging can produce exceptional images of tumor sites. These nanoparticles are much brighter than organic dyes and only need one light source for excitation which shows that the use of fluorescent quantum dots could produce a higher contrast image and at a lower cost than today's organic dyes used as contrast media. But the drawback is quantum dots are usually made up of quite toxic elements.

Another nanoproperty, high surface area to volume ratio, allows many functional groups to be attached to a nanoparticle, which can seek out and bind to certain tumor cells. Additionally, the small size of nanoparticles (10 to 100 nanometers), allows them to preferentially accumulate at tumor sites (because tumors lack an effective lymphatic drainage system).

Another use is with Sensor test chips containing thousands of nanowires, able to detect proteins and other biomarkers ("biomarker" refers to any of the body's molecules that can be measured to assess your health. Molecules can be obtained from your blood, body fluids, or tissue.) left behind by cancer cells, could enable the detection and diagnosis of cancer in the early stages from a few drops of a patient's blood.

The nanoshells can be targeted to bond to cancerous cells by conjugating antibodies or peptides to the nanoshell surface. By irradiating the area of the tumor with an infrared laser, which passes through flesh without heating it, the gold is heated sufficiently to cause death to the cancer cells.

Tracking movement can help determine how well drugs are being distributed or how substances are metabolized. It is difficult to track a small group of cells throughout the body, so scientists used to dye the cells. The way out of this problem is use of quantum dots, which is a tiny particle or nanoparticles of semiconductor materials (e.g. selenides or sulfide) of metals in all three spatial dimensions. They are more superior to traditional organic dyes. Quantum dots are 20 times brighter & 100 times more stable than traditional dyes.

Tissue engineering has been defined as "the application of principles and methods of engineering and life sciences towards fundamental understanding of structure-function relationships in normal and pathological mammalian tissues and the development of biological substitutes to restore, maintain or improve tissue function". The products that arise from these techniques may provide an alternative to available therapies to replace damaged, injured or missing body tissues.

Tissueengineered products (TEPs) typically are a combination of three components, i.e. isolated cells, an extracellular Matrix(all living things are made of cells but most of the cells in multicellular organisms are surrounded by a complex mixture of nonliving material that makes up the extracellular matrix -ECM) and signal molecules, such as growth factors. Nanotechnology provides new possibilities for the extracellular matrix, often referred to as the scaffold.

The extracellular matrix serves three primary roles. First, it facilitates the localisation and delivery of cells in the body. Second, it defines and maintains a three-dimensional space for the formation of new tissues with an appropriate structure. Third, it guides the development of new tissues with appropriate function. The interaction of the cells and the extracellular matrix is of great importance for the intended function of the final product. The excellent physical properties such as high surface area, high porosity, interconnective pores of the nanofibre matrices and appropriate mechanical properties, well-controlled biodegradation rates and biocompatibility, make (synthetic) biodegradable polymeric nanofibre matrices ideal candidates for developing scaffolds for TEPs, as reviewed by.

6. CONCLUSION

Nanotechnology influences almost every facet of life. When one goes down to the bottom of the material, one can discover unlimited possibilities and potential of the basic building block of the material (particle) which is different to that observed for the same material at bulk. The change in behaviour of material at nanoscale is dominated in the first place by quantum mechanics and is additionally attributable to material confinement in small space, and the increase in surface area per volume. At the nanoscale, physics, chemistry, biology, material science, and engineering converge toward the same principles and tools. As a result, progress in nanoscience has very far-reaching impact. Nanoparticles have potential applications in the field of medical sciences including new diagnostic tools, imaging agents & methods, targeted drug delivery, pharmaceuticals, bio implants and tissue engineering. Drugs with high toxic potential like cancer chemotherapeutic drugs can be given with better safety profile with the utility of nanotechnology. A single molecule of drug can be assisted to reach the desired site in order to reduce the side effects of the dose and its quantity. Quantum dots with MRI can produce excellent pictures of a tumor. Gold nanoshells can be used to detect, find, accumulate, and potentially destroy the tumor by heating the Nanoparticles. In the future, we can visualize a world with medical nanodevices, implanted or even injected into the body. A global perspective and collaboration might be needed in the field of research & development to give such benefits to mankind.

REFERENCES

- What is nanotechnology? https://www.nano.gov/nanotech-101/what/definition
- Merriam- Webster dictionary. https://www.merriam-webster.com/dictionary/nanotechnology.
- Oxford dictionaries. https://en.oxforddictionaries.com/definition/nanotechnology.
- Nanotechnology Definitions. https://www.understandingnano.com/nanotechnology-definition.html.
- Definition - What does Nanotechnology mean? https://www.techopedia.com/definition/3151/nanotechnology.
- Meet A. Moradiya, What is the Societal Impact of Nanotechnology. https://www.azonano.com/article.aspx?ArticleID=4992.
- Societal impact of nanotechnology. https://en.wikipedia.org/wiki/Societal_impact_of_nanotechnology.
- I.C. Gebeshuber, SOCIAL, HEALTH AND ETHICAL IMPLICATIONS OF NANOTECHNOLOGY. http://www.iap.tuwien.ac.at/~gebeshuber/Gebeshuber_Viennano07.pdf.

- Potential Impacts of Nanotechnology on Energy Transmission Applications and Needs. http://corridoreis.anl.gov/documents/docs/technical/APT_60861_EVS_TM_08_3.pdf.
- 3M, 2006, "3M's ACCR Overhead Conductor a Highlight of President Bush's Visit to Company's Labs; White House Interest Underscores Importance of Advance in Power Transmission," 3M News, Feb. 27. Available at http://solutions.3m.com/3MContentRetrievalAPI/BlobServlet?assetType=MMM_Image&locale=en_US&blobAttribute=ImageFile&fallback=true&univid=1114293973363&placeId=62603&version=current. Accessed July 27, 2006.
- Anderson, R., P. Chu, R. Oligney, R. Smalley, et al., 2006, White Paper, Smart Grid of the Future: A National Test Bed, Lamont-Doherty Earth Observatory, Columbia University. Available at http://www.ldeo.columbia.edu/res/pi/4d4/testbeds/Smart-Grid-White-Paper.pdf. Accessed July 7, 2006.
- Aspen, 2006, Aspen Aerogels. Available at http://www.aerogel.com/. Acc July 7, 2006.
- Davis, K., 2006, "Tiny Dreams for the Future of Transmission Capacity," Utility Automation & Engineering T&D Magazine, April. Available at

http://uaelp.pennnet.com/articles/article_display.cfm?article_id=253316&Section=ONART&C=INDUS. June 29, 2006.

- DOE (U.S. Department of Energy), 2006, "Cables and Conductors," Gridworks, U.S. Department of Energy, Office of Electricity Delivery and Energy Reliability. Available at http://www.energetics.com/gridworks/cables.html. July 13, 2006.

- EPRI (Electric Power Research Institute, Inc.), 2003, Electricity Technology Roadmap: 2003 Summary and Synthesis – Power Delivery and Markets, Nov. Available at http://www.hoffmanmarcom.com/docs/pd&m_roadmap_2003.pdf June 27, 2006.

- Foley, M., 2003, How Do Carbon Nanotubes Work: Carbon Nanotubes 101. Available at http://www.nanovip.com/node/2077. Accessed February 19, 2007.

- Foster, L., 2006, Nanotechnology: Science, Innovation, and Opportunity, Prentice Hall, Upper Saddle River, NJ.

- Fox, B., 2006, Efficiency Trials for Oxonica Nano Fuel Additive, Envirox. Available at http://www.azonano.com/details.asp?ArticleID=31. June 26, 2006.

- Gillett, S.L., 2002, Nanotechnology: Clean Energy and Resources for the Future, White Paper for Foresight Institute, Oct. Available at http://www.foresight.org/impact/whitepaper_illos_rev3.PDF. July 13, 2006.

- Gotcher, A., 2006, Written statement of Alan Gotcher, Ph.D., President and CEO, Altair Nanotechnologies, Inc., to the U.S. Senate Commerce Committee, June 14, 2006. Available at http://commerce.senate.gov/public/_files/Gotcher061406.pdf. July 7, 2006.

- Hoffert, M., 2004, "Renewable Energy Options – An Overview," in The 10-50 Solution: Technologies and Policies for a Low-Carbon Future, workshop cosponsored by the Pew Center on Global Climate Change and the National Commission on Energy Policy, Washington, D.C., March 25–26. Available at http://www.pewclimate.org/docUploads/10%2D50%5F Hoffert %2Epdf. February 19, 2007.

- IBM (IBM Business Consulting Services), 2004, "Revitalizing the Utilities Network," prepared for

Montgomery Research, March. Available at http://www-03.ibm.com/industries/.

- ROLE OF NANOTECHNOLOGY IN MEDICAL SCIENCES: A REVIEW. https://www.researchgate.net/publication/270684148_ROLE_OF_NANOTECHNOLOGY_IN_MEDICAL_SCIENCES_A_REVIEW.

www.ingramcontent.com/pod-product-compliance
Lightning Source LLC
Chambersburg PA
CBHW030700220526
45463CB00005B/1851